Connected,

or What It Means to Live
in the Network Society

Electronic Mediations

Katherine Hayles, Mark Poster, and Samuel Weber
SERIES EDITORS

Connected,

or What It Means to Live in the Network Society

Steven Shaviro

Electronic Mediations / VOLUME 9

UNIVERSITY OF MINNESOTA PRESS

MINNEAPOLIS • LONDON

Published by the University of Minnesota Press
111 Third Avenue South, Suite 290
Minneapolis, MN 55401-2520
http://www.upress.umn.edu

Library of Congress Cataloging-in-Publication Data

Shaviro, Steven.
　　Connected, or What it means to live in the network society / Steven Shaviro.
　　　　p.　　cm. — (Electronic mediations ; v. 9)
　　Includes bibliographical references and index.
　　ISBN 0-8166-4362-8 (cloth : alk. paper) — ISBN 0-8166-4363-6 (pbk. : alk. paper)
　　　　1. Information society.　2. Internet—Social aspects.　I. Title:
　　Connected.　II. Title: What it means to live in the network society.
　　III. Title.　IV. Series.
　　HM851 .S534　2003
　　303.48'33—dc21　　　　　　　　　　　　　　　　　　　2003009665

Printed in the United States of America on acid-free paper

The University of Minnesota is an equal-opportunity educator and employer.

12 11 10 09 08 07 06　　　　　　10 9 8 7 6 5 4 3 2

The boundary between science fiction and social reality is an optical illusion.

—DONNA HARAWAY

Each portion of matter may be conceived as a garden full of plants, and as a pond full of fish. But every branch of each plant, every member of each animal, and every drop of their liquid parts is itself likewise a similar garden or pond.

—G. W. LEIBNIZ

Angels are like eagles or tigers. They have no mercy, just a cold brilliance and glittering eyes watching for prey.

—MISHA

Everything has a schedule, if you can find out what it is.

—JOHN ASHBERY

The bell is the connection—which is more than junky-talk.

—JACK SPICER

Contents

Preface

Donna Haraway writes that in a world marked by rapid, startling innovations in information technology, electronic communications, and biological engineering, "the boundary between science fiction and social reality is an optical illusion" (149). Our lives are increasingly transformed in ways, and by devices, that seem to have come out of the pages of speculative fiction. At the same time, the largest tendency of these changes seems to be to transform the world itself into science fiction, or at least into a virtual-reality game. As Jean Baudrillard, among others, has remarked, under the reign of mass media and long-distance communications, reality itself has been turned into "its own pure simulacrum" (2001, 173). One need not share Baudrillard's Manichaeanism or his nostalgia for a supposedly lost Real in order to appreciate the cogency of his observations. Today, the technosphere, or the mediascape, is the only "nature" we know.

In this book, I try to write cultural theory as science fiction to come to grips with a world that itself seems on the verge of being absorbed into the play of science fiction novels and films. I have several precedents for this approach. In *Difference and Repetition,* Gilles Deleuze suggests that philosophy ought to be seen, in part, as "a kind of science fiction" (xx). This is because philosophy, like science fiction, can "make present the approach of a coherence that is no longer ours," no longer that of our familiar humanistic certainties. Philosophy is like science fiction in that it deals with concepts that have not yet been worked out; both genres work "at the border which separates our knowledge from our ignorance, and transforms the one into the other" (xxi). Or, as Michel Foucault similarly writes, the greatest reward and highest justification for intellectual work comes when this work results, "in one way or another and to the extent possible, in the knower's straying afield of himself" (1986b, 8).

In a different way, Carl Freedman argues, in his book *Critical Theory and Science Fiction*, that "theory" and "science fiction" have a privileged relationship. Both of these sorts of writing seek to grasp the social world not by representing it mimetically but by performing a kind of "cognitive estrangement" upon it (a term that Freedman borrows from Darko Suvin), so that the structures and assumptions that we take for granted, and that undergird our own social reality, may be seen in their full contingency and historicity. This means that science fiction is the privileged genre (literary, cinematic, televisual, and digital) for contemporary critical theory, in much the same way that the nineteenth-century realist novel was the privileged genre for the early-twentieth-century Marxist criticism of Georg Lukács and others. In addition, Freedman argues that critical theory and science fiction crucially share "certain *structural* affinities" (23) in the ways that they engage with late capitalist society. Science fiction and critical theory alike are engaged in the task of what Fredric Jameson calls the "cognitive mapping" of postmodern space: an effort that "seeks to endow the individual subject with some new heightened sense of its place in the global system," given that this system is unrepresentable by traditional mimetic means (54).

Connected, or What It Means to Live in the Network Society is a book about cyberculture, or about what Manuel Castells (2000b) calls the "network society." It is not an empirical study but rather a speculative exercise in cultural theory. As such, it is a work of science fiction in a way that I hope is consonant with both Deleuze and Freedman. Throughout this book I look at cultural practices, especially those involving digital media, both as they are described in science fiction novels and films and as they are being enacted today on the Internet. I do not distinguish between these two sorts of sources. My aim, like that of any other science fiction writer, is to discern the changes that are transforming our world into a very different place from the one into which I was born. Science fiction does not claim to predict what will happen ten, a hundred, or a thousand years from now; what distinguishes the genre is its linguistic and temporal orientation. Science fiction is always written in the future tense—conceptually, if not grammatically. Not only is it about what has not yet happened, but its very structure is that of the not-yet-happened. It addresses events in their potentiality, which is something vaster and more mysterious—more perturbingly *other*—than any actual outcome could ever be. Science fiction is about strange metamorphoses and venturesome, unpredictable results. It is a practice of continual experimentation, just as science and technology themselves are. In this way, science fiction conjures the invisible forces—technological, social, economic, affective, and political—that surround us. It makes those forces visible and palpable, and brings us face to face with them, however frightening and untoward they may be. It is only by writing cultural theory as science fiction that I can hope for my work to be (in Lenin's famous phrase) "as radical as reality itself."

Acknowledgments

I started working on this book because Jose Bragana de Miranda and Maria Teresa Cruz asked me to write something about networks for a conference they were organizing in Porto, Portugal, in 2001. The manuscript grew into far more than they originally asked for, but I must thank them for the initial impetus to gather together my eclectic thoughts and readings into a book.

Several people read portions of the manuscript as it was being written and offered me valuable feedback: Leo Daugherty, Lee Graham, Roddey Reid, Dominic Pettman, Richard Doyle, Carl Freedman, China Miéville, Samuel R. Delany, and above all Jacalyn Harden. When I presented parts of the book as talks, I received helpful comments from Doug Rice and his students at California State University at Sacramento, and from Friedrich Kittler.

I especially would like to thank Brian Massumi and Mark Poster for their warm encouragement of this project.

I finished writing this book just when my daughter, Adah Mozelle Shaviro, was born. This book about the uncertain future is for her.

Connected,

or What It Means to Live
in the Network Society

Only Connect. The word *connect* is an obscenity in the world of K. W. Jeter's science fiction novel *Noir*. People are always saying things like "connect you, mother-connector" (27) or "connect that" (192) or "get the connect outta here" (200). In short, if you're connected, you're fucked. Reach out and touch someone? It's the worst thing that could happen to you. Every connection has its price; the one thing you can be sure of is that, sooner or later, you will have to pay. The big problem today, we are told, is how to get everybody connected, how to get everybody onto the network. Our task is to overcome the digital divide, so that the wireless Internet is available to anyone, anywhere, at any time. This is supposed to be supremely democratic, not to mention an excellent marketing opportunity. But what do we really expect from such an intensive, twenty-four/seven connection? What is it that we really want? Jeter is not sanguine on this point. Every connection in *Noir* seems to lead back to the ubiquitous DynaZauber corporation. As Harrisch, a high-level DynaZauber executive, puts it: "In the marketplace, at least, rape is the natural order of things. And remarkably popular, too, on both sides of the exchange. People hand over their money, their lives, to DynaZauber or any other corporation, they know what they're getting. They want to get connected; the customers are always bottoms looking to get topped, the harder and bloodier, the better. That's the dirty little secret that corporations know" (314–15).

Song of the Jungle. Today, we are inclined to see nearly everything in terms of connections and networks. The network is the computer, we like to say. We think that intelligence is a distributed, networked phenomenon. A rain forest is an ecological network, according to both popular and scientific opinion. And, as Paulina Borsook points out, the technolibertarians of Silicon Valley and Redmond tend to regard the capitalist economy as a natural, organic network,

3

just like the rain forest (29ff). It's almost too perfect a meta-phor. The high-tech industry gets to have things both ways. On one hand, the rain forest is a place of life-and-death, Darwinian struggle. This is the famous vision of nature "red in tooth and claw," rape as the natural order of things, a war-rant for cutthroat capitalist competition. On the other hand, and at the very same time, the rain forest is a complex, self-regulating ecosystem. It exhibits spontaneous, self-generated order. All its pieces fit seamlessly together, and each prob-lem receives an optimal solution. Everything converges, as the New Economy evangelist Kevin Kelly puts it, into a universal, corporate "hive mind." The economy, like the rain forest, thus miraculously embodies both the New Age ideal of harmony and balance, and the workings of Adam Smith's invisible hand. All is for the best, in this best of all possible worlds, as long as nobody intervenes to limit cor-porate power. The economy as rain forest is a myth, in the precise sense defined by Lévi-Strauss: "a logical model ca-pable of overcoming a contradiction (an impossible achieve-ment if, as it happens, the contradiction is real)" (229). Such is the soft fascism of the corporate network: it reconciles the conflicting imperatives of aggressive predation on one hand, and unquestioning obedience and conformity on the other.

Stop the World, I Want to Get Off. For Jeter, the problem is not how to get onto the network, but how to get off. This is far more difficult than it might seem. For in-stance, you will never get television out of your life simply by turning it off and throwing away your set. It will follow you anyway, because the entire world exists only in order to be televised. A similar logic applies to the Internet. In an increasingly networked world, escape is nearly impossible. No matter what position you seek to occupy, that position will be located somewhere on the network's grid. No matter

what words you utter, those words will have been anticipated somewhere in the chains of discourse. As Burroughs
reminds us, "To speak is to lie — To live is to collaborate —."
There is no place of indemnity that would somehow be free
of these constraints. Nonetheless, Burroughs continues,
"there are degrees of lying collaboration and cowardice —
...It is precisely a question of regulation —" (1992b, 7).
You cannot opt out of the network entirely, but at the very
least, you can try to be connected a little less. You can provide your own negative feedback. You can regulate your
own contributions to the system that is regulating you.
What's needed for this, no doubt, is a kind of ironically distanced, self-conscious asceticism. The insidious thing about
electronic networks is that they are always there, whether
you pay attention to them or not. Indeed, they assume, and
even require, a kind of distracted inattention on your part.
You can never directly confront the network, stare it straight
in the eye. For it is always somewhere else from wherever
you may be looking. But such enforced distraction can also
be cultivated for its own sake. And in this way, perhaps,
distraction might become a space in which to breathe.

Medium Cool. Marshall McLuhan famously argued that
television is a cool medium (1994, 22–32 and 308–37). It does
not try to shock and overwhelm us, the way that movies
seen in theaters do. Rather, TV is laid back and low intensity. It's not a visual medium, McLuhan says, so much as
an aural-tactile one. The TV image "is not photo in any sense,
but a ceaselessly forming contour of things limned by the
scanning-finger...the image so formed has the quality of
sculpture and icon, rather than of picture" (313). In television, Michel Chion similarly points out, "sound, mainly
the sound of speech, is always foremost," and the image is
just "something extra" (157–59). Voices provide continuity,
while the images continually change. This is the opposite

situation from film, which is anchored in the image. TV, unlike cinema, is intimate and close range. It is really just part of the furniture. Often we leave it on in the background, as we go about our daily chores. But even when we pay it close attention, it does not stupefy us and make us passive. To the contrary, it invites our participation. We channel surf, we make snide remarks, we yell back at the set. It's silly to think that anyone is brainwashed by TV. It doesn't constrain us, or perpetrate violence upon us. Much more subtly and insidiously, TV draws us into discourse, absorbs us into the network. It colonizes us obliquely, by distraction. It allures us, willy-nilly, into getting connected. We may say of television what Foucault says more generally about postmodern power: it doesn't constrain us or repress us, so much as "it incites, it induces, it seduces" (1983, 220). It persuades us or cajoles us into doing the work of policing ourselves. As Harrisch explains toward the end of *Noir*, the true purpose of the network is "the translation into reality of all those Foucauldian theories of self-surveillance. The brain watches itself and administers its own stimuli and rewards, with DynaZauber as the beneficiary" (463).

The Body and the Screen. The Internet is even cooler than television. That is to say, it is even lower definition than TV and, consequently, even more involving. The World Wide Web offers possibilities so vast, and yet so tantalizingly incomplete, that I must get involved with it in depth. I am drawn in, I can't help myself. This is why the Net is an interactive, many-to-many medium, whereas TV is only one-to-many. Television addresses my ears and eyes, but the Net solicits my entire body. Web surfing is a tactile, physical experience. In the first place, it requires the correct posture. I must sit upright, directly in front of the screen, without slouching, and with my arms horizontal and my hands engaged. I must also remain much closer to the screen than ever is the case with TV, close enough to read the small print and to watch the jerky video clips that run in postage-stamp-sized windows. Meanwhile my fingers are running across the keyboard. My right hand keeps busy moving and clicking the mouse. In this way, the hand becomes an extension of the eye: I reach right into the screen and travel through its iconic, hyperlinked space. Cyberspace is what Deleuze and Guattari call a "haptic" space, as opposed to an optical one: a space of "pure connection," accessible only to "close-range vision," and having to be navigated "step by step.... One never sees from a distance in a space of this kind, nor does one see it from a distance" (1987, 492–93). No panoramic view is possible, for the space is always folding, dividing, expanding, and contracting. Time is flexible on the Net as well; things happen at different speeds. Sometimes I must read and type extremely fast to keep up with rapid-fire chat room conversations. Other times I have to hold myself back as I wait for pages or files to download. What's more, these multiple speeds, times, and spaces overlap. Enveloped in the network, I am continually being distracted. I can no longer concentrate on just one thing at a time. My body is pulled in several directions at once, dancing to many distinct rhythms. My attention fragments and multiplies as I shift among the many windows on my screen. Being online always means multitasking.

Distraction. Bruce Sterling's science fiction novel *Distraction* deals with, among other things, the phenomenology of multitasking. Oscar Valparaiso, the protagonist, is infected with a virus that multiplies his awareness. It modifies his brain in such a way that he develops two separate centers of consciousness. He is able to pay attention to two different things at once, instead of having to move sequentially between one and the other. Oscar has "two windows open on the screen behind his eyes," so that he is literally "multi-tasking, but with his own brain" (438). Such a double consciousness is inherently paradoxical. There is more than one subject in Oscar's mind that says "I." Yet these are not multiple personalities, since Oscar is the same person for both. Nor do they reflect a primordial split, as with the Freudian unconscious or the Lacanian Other. For both centers are entirely self-present and conscious. Oscar experiences no lack; if anything, his brain is full to bursting: "Oscar could actually feel the sensation, somatically. It was as if his over-tight skull had a pair of bladders stuffed inside, liquid and squashy, like a pair of nested yin-yangs" (496–97). No Cartesian dualism here: thought is a visceral experience. Oscar also finds that he is always murmuring to himself; this is how his attention centers inform each other of what they are doing. When consciousness multiplies in this manner, there can be no problem of "other minds" or of what Wittgenstein called "private languages." For everything is intelligible, and nothing is hidden. Oscar's brain is an open network, with massively parallel processing. Such mental networking is the source of what we commonly call intuition and creativity; as someone tells Oscar, "when you're really bearing down, and you're thinking two things at once—ideas bleed over. They mix. They flavor each other. They cook down real rich and fine. That's inspiration. It's the finest mental sensation you'll ever have" (513). But Oscar pays a price for such mental agility: he finds that he suffers from "poor impulse control" (513). The problem is, he can't hold anything back. The ideas come thick and fast, and they demand immediate expression. The clamor in his brain leaks out into the world. When attention is so magnified, multiplied, and multitasked, the result is indistinguishable from complete distraction. It's like having every available television channel on at the same time.

Come to Daddy. Chris Cunningham renders the creepy terror of the network in his music video for the song "Come to Daddy" by Aphex Twin, aka Richard D. James. The song is a piss-take on heavy metal: five and a half minutes of screaming Satanic fury ("I want your soul"), set against synthesized guitar fuzz, electronic bleeps, and an energetic drum 'n' bass backbeat. The sound is heavy and violent, but at the same time static and synthetic. The identity of visceral intensity and ostentatious fakeness is, of course, the point of James's joke. The video is a miniature horror film, set in a ravaged urban landscape in front of a row of high-rise slums. Feral children run wild in the streets. They bang on fences, smash cars, hurl garbage cans, and terrorize passersby. Even more disturbingly, these children, regardless of gender, all have James's obscenely grinning face. A TV set abandoned in the gutter suddenly comes to life when a dog pisses on it. A demon screams and snarls on the tube. Eventually, the demon oozes out of the set and materializes on the street, in a scene cribbed from David Cronenberg's *Videodrome*. The image becomes physical. Cunningham's editing matches the music's stuttering rhythms, alternately stretching and compressing time. The video cuts back and forth between close-ups and long shots, as between images on the television screen itself, and images of the wasteland in which the set was found. Patterns of shot and reverse shot do not map out consistent spatial relations; rather they work as relays, creating a dense network that, in the words of Deleuze and Guattari, "connects any point to any other point" (1987, 21). The video refuses to distinguish between physical space and screen space, or between actual objects and their virtual, fictive representations. Everything is both body and image, and every body/image has the same ontological status. There is no privileged spectatorial perspective and no distance between the spectator and the screen. There is only the video network; along its pathways, the nightmarish image of Richard D. James's face proliferates indefinitely, like a virus.

The Integrated Circuit. Our current understanding of networks dates from the development of cybernetic theory in the 1940s and 1950s. The model has since been greatly elaborated, notably in the chaos and complexity theories of the 1980s and 1990s.[1] As it seems to us now, a network is a self-generating, self-organizing, self-sustaining system. It works through multiple feedback loops. These loops allow the system to monitor and modulate its own performance continually and thereby maintain a state of homeostatic equilibrium. At the same time, these feedback loops induce effects of interference, amplification, and resonance. And such effects permit the system to grow, both in size and in complexity. Beyond this, a network is always nested in a hierarchy. From the inside, it seems to be entirely self-contained, but from the outside, it turns out to be part of a still larger network. A network is what Ilya Prigogine and Isabelle Stengers call a dissipative structure: it generates local stability and maintains internal homeostasis in far-from-equilibrium conditions, thanks to massive "energy exchanges with the outside world" (143). These expenditures, in turn, become a source of hidden order, what Stuart Kauffman calls "order for free." What seems like noise, waste, chance, or mere redundancy from the point of view of any given system turns out to be meaningful and functional in the context of the next, higher-level system. As N. Katherine Hayles puts it, networks operate through a dialectic of pattern and randomness, rather than one of presence and absence (285). This is why we define networks as being made of bits, rather than atoms. What matters is not the hardware, but the software; not what the network is actually made of, but only the way it is connected up, and the information that gets transmitted through it.

The Algebra of Need. In *Naked Lunch,* Burroughs describes the "basic formula" of addiction, or what he calls the algebra of need: "The face of 'evil' is always the face of total need . . . Beyond a certain frequency need knows absolutely no limit or control." The addict has no choice; she or he will "do anything to satisfy total need" (xi). Of course, Burroughs's formula does not just apply to heroin addiction; it is a crucial component of the power relations that undergird all of consumer culture. Total need guarantees total participation; and total participation means total subjection. A perfect commodity is anything that—like junk for the junkie—I absolutely have to buy. Now, there is a profound synergy between the algebra of need and the logic of self-regulating networks. There are two reasons for this. First, the very structure of the network works to perpetuate infinite need. A network has no goal outside itself and therefore no objective measure of satiety. It strives only to maintain itself. But such self-maintenance is an endless task: a network's labor is never done. A network must make vast expenditures, simply in order to preserve, or reestablish, homeostasis. It's just like the later stages of heroin addiction. After a while, the drug doesn't get you high any longer; you need an ever-larger dose merely to get straight. As Jeter puts it in *Noir,* with explicit reference to Burroughs, at this point you will "pay any price just to feel back to normal again" (460). A network, just like an individual junkie, gives out more and more over time, only to receive less and less back. In the second place, the logic of networks tends toward the algebra of need because the addiction process is facilitated and accelerated when materiality is replaced by information. Information is a lot cheaper to produce than physical goods. It can be transported much more easily. It's nearly an ideal commodity because it feeds back upon itself and incites its own want: the more of it you have, the more you need. Also, information can be replenished indefinitely by the producer and sold to more and more customers at almost no additional cost. From the point of view of its corporate suppliers, information indeed wants to be free. But for the consumer, information remains a scarce resource because even too much is never enough. Jeter suggests that this prospect of generating higher revenues at lower costs is the real reason for "the whole push to get people on the telecommunications wire, have them value bits of information as much or more than the atoms of the real world, have them pay to be mesmerized by the pretty color lights on their computer screens" (461).

Fractals and Viruses. The network is shaped like a fractal. That is to say, it is self-similar across all scales, no matter how far down you go. Any portion of the network has the same structure as the network as a whole. Neurons connect with each other across synapses in much the same way that Web sites are linked on the World Wide Web. McLuhan claims that "electronic circuitry [is] an extension of the central nervous system" (1967, 40). But the opposite formulation may well be more useful: every individual brain is a miniaturized replica of the global communications network. The network is the great Outside that always surrounds and envelops me. But it is also the Inside: its alien circuitry is what I find when I look deeply within myself. The network is impersonal, universal, without a center, but it is also perturbingly intimate, uncannily close at hand. This is why Deleuze defines subjectivity as a *fold:* it is "an interiorization of the outside...a redoubling of the Other...a repetition of the Different...It resembles exactly the invagination of a tissue in embryology" (1988a, 98).

Burroughs makes a similar point when he suggests that "the whole quality of human consciousness, as expressed in male and female, is basically a virus mechanism" (1981, 25). In both cases, identity is implanted in me from without, not generated from within. My selfhood is an information pattern, rather than a material substance. I may describe this process that subtends my consciousness in several ways: as embryonic infolding, as fractal self-similarity, or as viral, cancerous proliferation. But the difference between these alternatives is just a matter of degree. The crucial point is that the network induces mass replication on a miniaturized scale and that I myself am only an effect of this miniaturizing process.

Viral Marketing. During the Internet business bubble of the late 1990s, there was much talk of *viral marketing.* In 1999, a venture capitalist even noted that "almost 80% of the business plans we see have the word 'viral' in them" (Sansoni). What lay behind such a fad? Viral marketing is defined, on Whatis.com, as "any marketing technique that induces Web sites or users to pass on a marketing message to other sites or users, creating a potentially exponential growth in the message's visibility and effect." The idea is to send out an advertisement that "spreads like a disease" (Willmott). The message propagates itself by massive self-replication as it passes from person to person in the manner of an epidemic contagion. This is supposed to be more than just a metaphor. The viral message is composed of memes in the same way that a biological virus is composed of genes. The memes, like the genes, enter into a host and manipulate that host into manufacturing and propagating more copies of themselves. Packages of information spread and multiply, just like packages of DNA or RNA. And the mathematical model of how this process works is precisely the same for information viruses as it is for biological viruses.

If viral marketing works at all, it is quite an efficient strategy. The message is propagated at an exponentially expanding rate as it follows the fractal shape of the network. You get wide distribution at a low price because "customers do the selling" (Jurvetson and Draper). That is to say, the cost of manipulating and exploiting the consumer is offloaded onto the consumer herself. It's even better than having a slave who works for free: the customer actually pays for the opportunity to be your shill. This sort of viral infection, and interiorization, is the final step in the algebra of need: now the entire network is miniaturized and implanted directly into the brain of every individual consumer. Jeter imagines a hyperbolic form of this operation in *Noir*. Harrisch rhapsodizes that the process is "the height of mercantile capitalism: you chain your customers to their lathes and running-shoe assembly lines, and you throw the key to the padlock down the black hole you've put inside their heads... the whole thing runs itself... What could be more beautiful than that?" (463).

The Selfish Meme. How seriously can we take Richard Dawkins's meme hypothesis (1989, 189–201)? Dawkins broadly defines the meme, by analogy with the gene, as "a unit of cultural transmission, or a unit of imitation... Examples of memes are tunes, ideas, catch-phrases, clothes fashions, ways of making pots or of building arches" (192). Dawkins proposes this concept in order to suggest that the logic of Darwinian natural selection, "the differential survival of replicating entities" (192), might well be at work in other media than DNA base pairs (for instance, in connections among neurons), resulting in the evolution of forms other than biological organisms (such as ideologies, religions, and fads). Memes have become a hot topic in recent years. But the deeper implications of the theory have been ignored both by Dawkins's opponents (Lewontin et al.)

and by the enthusiasts of the supposed science of *memetics* (Lycaeum). The meme hypothesis is actually quite close (closer, I imagine, than Dawkins would be comfortable with) to Burroughs's account of consciousness and language as viral mechanisms. If we take seriously the idea that memes, like genes, are always engaged in a Darwinian struggle to survive and reproduce, then we cannot assimilate them to the fashionable view that the world is nothing but patterns of information. Dawkins implies as much with his notorious, wonderfully lurid, description of multicelled organisms, ourselves included, as "gigantic lumbering robots" created by the genes within them as "survival machines" (19). This statement is often vilified, with much self-righteous humanistic indignation, as a declaration of DNA determinism. But contrary to this misreading, Dawkins's formulation actually insists on the impossibility of any unilateral control by the genes. It suggests, rather, an almost Manichaean duality. There's an irreducible gap between replicator and vehicle, between genotype and phenotype, between software instructions and hardware implementation: in short, between the ideality of a repeating informational pattern, and the contingency of any particular material embodiment. Whether it is a question of genes or memes, DNA or ideas, the processes of transcription and implementation are never smooth and transparent. "Gigantic lumbering robots," after all, are crude and clumsy. They do not follow instructions easily or accurately. What's more, there is no reason why the robot's own interests should coincide with those of its controllers. More than likely, these interests will actively clash. The Darwinian war between rival genes or memes is also a war between all these replicators, on the one hand, and the vehicles that bear them, on the other. This is why Dawkins ends his discussion of memes with a call for us to "rebel against the tyranny of the selfish replicators" (201).

A Universal History of Parasitism. A pattern of information is meaningless by itself. A virus remains inert unless it encounters a suitable living cell. A configuration of ones and zeroes is similarly no more than gibberish until it is processed by the right program. Genes and memes are helpless without their hosts. They need to be instantiated in flesh, or at least in matter. They can only replicate themselves by means of the effects they have on bodies. But these effects are multiple, contradictory, widespread, and often indirect. We cannot think of information as just a pattern imprinted indifferently in one or another physical medium. For information is also an event. It isn't just the content of a given message but all the things that happen when the message gets transmitted. As Morse Peckham puts it, "the meaning of a verbal event is *any* response to that event" (1988, 16; emphasis added). In other words, meaning is not intrinsic, but always contingent and performative. "By their symptoms you shall know them," Burroughs reminds us; "if a virus produces no symptoms, then we have no way of knowing that it exists" (1981, 24). Dawkins explores the furthest reaches of such a symptomatology in his book *The Extended Phenotype*. To assess a strand of DNA, we must look at the full range of responses it elicits. A gene does not just affect the shape and behavior of the body within which it happens to be enclosed. It also influences other organisms (as when a parasite manipulates the behavior of its host), the physical landscape (as when beavers build dams), and the atmosphere (as when plants pump out oxygen). In theory, such "genetic action at a distance" has no limit; the entire "living world can be seen as a network of interlocking fields of replicator power" (247). The individual organism is only a transitory by-product of the multiple processes running through this network. I cannot make any categorical distinction between the replicators (genes or memes) that would intrinsically belong to me and those (biological or linguistic viruses) that would have only affected me from outside. Over the course of time, either of these may turn into the other. And in any case, there always remains a gap between the genotypic instructions themselves and their phenotypic inscription and implementation in my flesh. The imperatives that give rise to me are never truly my own. Parasitism thus becomes a universal principle.

Bloodchild. Octavia Butler's science fiction short story "Bloodchild" is about the ambivalent relations between a parasite and its host. Butler describes the tale as a "pregnant man story" (30). In "Bloodchild," human beings live precariously on a planet dominated by the Tlic, who are sort of like three-meter-long, intelligent centipedes. In order to reproduce, the Tlic must lay their eggs in the viscera of a warm-blooded animal. The eggs are nourished by the host's blood. Larval grubs eventually emerge; they burrow and eat their way through the host's body. Human males are the hosts of preference. Gan, the story's narrator, is one such host. He is groomed for his role from early childhood and develops a close relationship with T'Gatoi, his future inovulator. Gan tells us of his doubts and hesitations, and he spares us none of the gruesome details of the incubation process. But in the end, he submits to his preordained role. What happens is not exactly a rape, but it *is* a kind of emotional blackmail. For T'Gatoi wants not just Gan's body, but his consent to the procedure as well. And she gets it. Gan accepts his own subordination. He realizes that he loves T'Gatoi, despite (and because of) their enormous differences from one another and the gross inequality between them. Loving somebody means allowing that body into the deepest recesses of yourself. It means permitting the one you love to use you, to have power over you. For power, as Foucault says, is neither violent coercion, nor the free accord of equals. It is rather "a mode of action upon the action of others," which means that "it is exercised only over free subjects" (1983, 221). This explains why Butler denies that "Bloodchild" is about slavery (30). The story is so disturbing precisely because Gan is not a slave; he willingly embraces his fate of being a host for Tlic parasites. "Bloodchild" raises all sorts of questions about otherness and difference; about cultural norms and gender roles and conceptualizations of race; and about power, love, and vulnerability. But it also demands that we take it literally. The story's invented biology is not just a metaphor. The Tlic do not stand in for anything human. They remain profoundly alien, in body and in mind. Gan may love T'Gatoi and offer his body to her; he knows, too, that she chose him and loves him in her own way. But there is no true reciprocity between them; their bodies, their needs, their feelings, and their powers are far too different. The parasite will never fuse with its host in blessed, perfect symbiosis.

My Vocabulary Did This to Me. For Jack Spicer (1998), every poem is a parasite, and the poet is its host. Writing poetry is literally a matter of taking "dictation." The poet receives messages like a radio and transcribes them like a medium at a séance. Poetry, therefore, is not self-expression. Rather, you know that you have produced a poem, precisely when the poem says "the thing that you didn't want to say in terms of your own ego, in terms of your image, in terms of your life, in terms of everything" (6). A poem is a message, but one that doesn't really concern us. It is sent by forces or entities unknown, and it is not addressed to the poet, nor even necessarily to the reader. The important thing to realize, Spicer says, "is essentially that there is an Outside to the poet" (5). The poem arises out of this insurpassable gap, this "absolute distinction between the Outside and the inside" (7). Spicer does not know what lies Outside. The entities who dictate his poetry are not sources of truth or higher wisdom. They are more like Martians in 1950s science fiction movies, or even the early-1960s sitcom figure "My Favorite Martian": sinister or funny aliens with their own odd purposes. These Martians do whatever they do "for pretty damn selfish reasons" (34). They are brain parasites: they enter your mind and take whatever "furniture" they find there—"your memories, your language"— and "rearrange" this furniture like "alphabet blocks" to spell out their messages (9, 24). The poem is written in the poet's own language; it is made out of his or her own memories, ideas, desires, and obsessions. But the message composed with these materials is not anything the poet wants to say. What speaks in the poem? Not memory, not desire, and not even language. Although the idea of "poems following the dictation of language" was popular throughout the twentieth century, Spicer insists that it is "nonsense" (9). Whatever speaks in the poem is radically other. It is refractory to my language, to my memories, and to my desires. Yet it is compelled to make use of all these things as tools to send its message out. The poet's task is to channel the otherness, to clear a space for it. I must welcome the parasite into my own brain. But no matter what happens, I will never be more than a transitory host. I can never appropriate this otherness for myself.

Language Is a Virus. The Beat poet Lew Welch (1926–1971) never achieved the same degree of recognition as his friends and fellow writers Gary Snyder, Philip Whalen, and Jack Kerouac. Not many people read Welch's poetry during his own lifetime; even fewer people know it today. Nonetheless, Welch wrote one line that became more famous than anything written by any of his contemporaries. For many years, everyone in America was familiar with it. Alas, the line was not from any of Welch's poems. Rather, it was a slogan that he wrote for his day job with an advertising agency, and that was incessantly used in TV and print commercials. The line was this: "Raid Kills Bugs Dead" (Saroyan).

Everything Is True, Everything Is Permitted.
Burroughs describes the six ancient Cities of the Red Night. These cities are way stations in the journey of the human soul. But they also stand for different types of social organization. One of them is Ba'dan, a place "given over to competitive games and commerce ... Unstable, explosive, and swept by whirlwind riots." It has "a precarious moneyed elite, a large disaffected middle class and an equally large segment of criminals and outlaws." And its ruling ethos is that of the free market, where "everything is true and everything is permitted." In short, Ba'dan "closely resembles present-day America" (158). Burroughs contrasts the continual ferment of "creative destruction"[2] in Ba'dan with the harsh regimentation of life in Yass-Waddah, a religious fundamentalist city where "everything is true and nothing is permitted except to the permitters" (158). Ba'dan and Yass-Waddah face one another across a river (153), and they are locked in perpetual conflict. For they represent the two faces of the American Dream, or what Harold Bloom calls the "American Religion." When white people first invaded North America, they were motivated by both greed and bigotry. They were libertarian individualists to whom everything was permitted; they were also evangelical zealots, determined to impose their system of prohibitions upon everybody else. Both attitudes have their roots in the same White Anglo-Saxon Protestant sense of absolute entitlement. For Americans, "everything is true"; like Mulder in *The X-Files*, we "want to believe." Now, Burroughs was a full-fledged partisan in the culture wars of the late twentieth century. He urges "an all-out assault on Yass-Waddah ... Even the memory of Yass-Waddah must be destroyed as if

Yass-Waddah had never existed" (284). But he also knows
that such a war is unwinnable, if its only aim is to restore
the traditional privileges of Ba'dan, for the two cities are
closely intertwined and mutually dependent. On the other
hand, a great distance separates pragmatic and reductionist Ba'dan, with its claim to every truth, from the two final
Cities of the Red Night: the "cities of illusion" Naufana
and Ghadis, where—in accordance with Hassan I Sabbah's
maxim—"nothing is true and therefore everything is permitted" (159).

Possible Worlds. In his "Fall Revolution" series of science fiction novels, Ken MacLeod looks at the relations
between technology, on the one hand, and political and economic organization, on the other. The first book, *The Star
Fraction,* sets forth the main premises of the series. The network is never neutral and never merely technical. Rather, it
is political to the core. Its development is largely driven by
economic and ideological forces. It is both a weapon and a
stake in ongoing political struggles. Trotskyist militants,
free-market libertarians, and fundamentalist Christians all
use the network; it is continually being reshaped by their
conflicting demands. The other volumes in the series offer
contrasting perspectives upon these conflicts. The second
book, *The Stone Canal,* is set in a libertarian capitalist world.
The third novel, *The Cassini Division,* envisions a socialist
utopia. And the final volume, *The Sky Road* , takes place in
a Green society that severely limits technological development. These worlds have radically different assumptions
and expectations. Even when they use the same material
technologies, they set them up in sharply divergent ways.
Each society, therefore, provides divergent perspectives
upon the others. The socialist Ellen Ngwethu, for instance,
is shocked by the "dazzle of clashing colours and . . . blare of
sound" that assaults her when she visits the ultra-capitalistic

planet of New Mars: "Every square yard was occupied by a stall or shop or kiosk, each of which had its own fluorescent rectangle above it advertising flights or drugs or socks or cosmetics or lingerie or insurance or back-ups or cabs or hotels. The public-address system thumped out urgent-sounding music made all the more unsettling by frequent, and equally urgent-sounding, interruptions" (1999, 183–84). Now, Ellen herself is no puritan; her own world is a high-tech, heavily networked cooperative commonwealth, characterized by material abundance and hedonistic mores. Yet nothing in her world has prepared her for a scene like this. Among so many distracting, demanding choices, she has no idea even where to begin. MacLeod is showing us our own market-crazed society, of course, as it might appear from the other side of the mirror.

Media Saturation. The science fiction comic book series *Transmetropolitan* is written by Warren Ellis and drawn by Darick Robertson. The series is set in a future American mega-metropolis known only as the City. It's a place much in the mold of Burroughs's Ba'dan. Every square inch of the City is filled with crowds in ceaseless motion, continually being assaulted by flashing billboards, multimedia screens on the sidewalk, surveillance cameras, and street vendors selling everything from high-tech sex toys to "long pig," fast food made from cloned human flesh. People alter their genetic makeup at the slightest whim; with "temporary morpho-genetic plug-ins," for instance, you can be a tourist in another species, "get yourself some reptile skin for three weeks, spend a month in feathers" (1998a, 37). You can also imbibe "gene factors" that plug you permanently into the worldwide communications network or that cure whatever new diseases have invaded the City lately. People sometimes even download themselves into "foglets," clusters of airborne nanomachines that take the place of more palpable flesh (1998b, 72–92). But if you don't want to go that far, you can try the new drugs that are constantly being synthesized, faster than the government can outlaw them. Or else, as a last resort, you can convert to one of the new religions that are being invented daily (1998b, 48–71). And if it's self-protection you're after, there are lots of bizarre weapons around, such as the dreaded "bowel disruptor" (1998b, 24). Of course, there is no such thing as privacy in the City. The media are everywhere. Thousands of TV channels cater to every imaginable age, ideology, taste, and sexual preference. Cameras, microphones, and reporters are stationed on every block, providing live feeds to the Net. In short, the delirium of advanced technology has been entirely woven into the texture of everyday life. This is what a fully networked, "posthuman" existence might be like. Could we really endure such a condition? One episode of *Transmetropolitan* is about Revivals, people who had their bodies placed in cryogenic suspension when they died at the end of the twentieth century (1998b, 93–114). Brought back to life in the City, these people immediately freak out. They simply can't handle the overload, and nobody much wants them around, anyway.

The Message Is the Medium. The messages assault us, from within and from without. In *Noir*, Jeter shows us corporate executives with "swarms of E-mail buzzing around their heads ... tiny holo'd images yattering around them for attention" (18). And similarly, in his novel *Nymphomation*, Jeff Noon writes almost gleefully of "blurbflies," artificial insects whose buzzing songs transmit advertising messages. When such insect messengers come calling, you cannot choose not to respond. You may swat or shoo away a single fly, but more of them will always show up. The mediasphere is the only "nature" we know, and unwanted messages, like insect pests, are a crucial part of its "ecology."

I do not like the incessant barrage of ads on TV, on the Web, and in my e-mail, any more than I like mosquitoes and bedbugs, but I know that the network is inconceivable without them. It has been estimated that, in the biological realm, eighty per cent of all organisms are parasites (Zimmer xxi). An even higher percentage of the human genome seems to be nothing but "junk DNA" (Dawkins 1998, 97ff). Why should the mediasphere be any different? To say that this is all just "information" misses the point. For such a description ignores the fact that so many of the messages we receive are aggressive or hostile, when they aren't simply indifferent to us. A message is not just passively conveyed from the sender to the receiver through a pre-existing medium. We should rather say, inverting McLuhan's dictum, that the message is itself the medium. Those blurbflies are not just accidental features of an otherwise neutral environment; they themselves *are* the environment that we live in. Foucault argued that there is no such thing as Language; there is only a multitude of particular statements, each of which is more an action than a signification. The accumulation of these statements over time leaves behind traces and contours that we retrospectively construe as "regularities of discourse."[3] Similarly, the network is not a disembodied information pattern nor a system of frictionless pathways over which any message whatsoever can be neutrally conveyed. Rather, the force of all the messages, as they accrete over time, determines the very shape of the network. The meaning of a message cannot be isolated from its mode of propagation, from the way that it harasses me, attacks me, or parasitically invades me.

Buy Me. In *Transmetropolitan,* we are introduced to a hot new marketing phenomenon: "block consumer incentive bursting," popularly known as "buybombs." A flash of light zaps you from the television set. You see intense patterns and spots before your eyes; pressure from the flash can even give you a nosebleed. But the major effect only comes later, when you go to sleep. The buybombs "load your brain with compressed ads that unreel into your dreams" so that you are literally "dreaming television advertisements" (Ellis and Robertson 1998b, 45–47). This is viral marketing with a vengeance. What better tactic, for memetic parasites, than to infiltrate themselves into our dreams? It's not just that dreams are so common a metaphor for our most fervent hopes, desires, and imaginings, nor even that we find it so easy to believe that they provide us with insight into ourselves. The really important thing about dreaming is this: it is the most antisocial activity I ever engage in. Dreaming is the one experience that I must go through alone, that I cannot possibly share with anyone else. Of course, I often try to recount my dreams in words, but I cannot help feeling that such words are woefully inadequate. If I cannot quite remember my own dreams, if I cannot narrate them to myself any more than I can to someone else, if I cannot translate them into the generalities of language, this only confirms how unique and personal they are. Dreams are the last refuge of old-fashioned interiority and mental privacy. This makes them the object of a powerful nostalgia, even among the most resolutely unsentimental and forward-looking. My dreams are so many proofs of my singularity; they cannot be redeemed, or substituted for, or exchanged. That's why any violation of dream space is so disturbing, whether it's Freddie Krueger assaulting me in my sleep or buybombs exploding in my head. It means that I haven't really withdrawn from the world after all. It means that I am nothing special. It means that I'm just the same as everybody else. The network has colonized my unconscious. It has made me into a tiny version of itself. The whole situation reminds me of Jim Carrey's Ace Ventura, spewing out forty years' worth of television impersonations as if they were the raw contents of his unbridled id. Such is the terminal state of the networked consumer: to be intensely involved, and maximally distracted, all at once.

Arms Race. In *Noir,* Jeter argues that advertising revenues will never be sufficient to pay the costs of the network. As ads become more numerous and more obtrusive, he suggests, people will discover new ways to resist and ignore them. By the time of the future world of *Noir,* "the evolution of the human brain had taken care of the situation. A filter is like an immune system, and vice versa; it hadn't been too long before a benign mental cataract had been determined to exist, one that spread memelike through the species. Ad-blindness, linked to the refresh rates on visual display units" (252–53). Of course, this victory is unlikely to be the end of the story. The struggle between advertisers and consumers bears all the hallmarks of an evolutionary arms race. "There are arms races between predators and prey, parasites and hosts," as Richard Dawkins explains it. "They consist of the improvement in one lineage's (say prey animals') equipment to survive, as a direct consequence of improvement in another (say predators') lineage's evolving equipment" (1987, 178). Frequently these arms races spiral out of control, leading to what biologists call (after Lewis Carroll) the "Red Queen effect": both sides are forced to run faster and faster, just to keep in the same place relative to one another (183ff). Already, today, "software that stops advertisements from appearing on screens is free to any home user who chooses to download it"; in response, Internet advertising companies are beginning to "develop the technology to 'find a way around' programs that detect ads" (Stamler). The competition will only get more intense, especially when the arena of conflict moves (as Jeter suggests it will) from software on our computers into our actual neurons and synapses. It's the twenty-first century extension of a process already described by Walter Benjamin, with regard to cinema, in the early twentieth: "the spectator's process of association in view of these images is indeed interrupted by their constant, sudden change. This constitutes the shock effect of the film, which, like all shocks, should be cushioned by heightened presence of mind" (1969, 238). In the age of cinematic, mechanical reproduction, filmmakers were always upping the ante, in terms of the shocks with which they assaulted the audience. In turn, audiences responded to these shocks by becoming habituated to them and by cultivating an ever greater sense of distraction (240–41). In our current, cooler and more intimate, media atmosphere, the process is more one of insinuation than of shock: the ads irritate me like an itch on an intimate body part that I am unable to scratch. But the play of violation and defense, or of aggression and distraction, as they match one another in an ever escalating arms race, remains pretty much the same.

Rumble in the Bronx. Mike Ladd's experimental hip-hop album, *Gun Hill Road* by the Infesticons, tells the story of an epic struggle on the streets of the North Bronx. The down 'n' dirty Infesticons fight to save the world from the Majesticons, robots with a mission to "jiggify" everything. The Infesticons are Spartan and austere, "interested in ideas and the content of their minds." The Majesticons, to the contrary, seem to be a stand-in for Sean "Puffy" Combs (now known as P. Diddy) and his crew. They love glitz and glamour and "pink cocaine in a cognac glass"; they are "fascinated with their own exterior, and [develop] a hierarchy based on one's sense of style." The war between these rivals is fought on the plane of sound. The Infesticons' music is dissonant and distorted: dense, rapid-fire rants, laden with twisted rhymes and puns, set against screaming psychedelic guitars, electronic squiggles, and sampled industrial noise, to a shifting background of heavy funk beats. The Majesticons' music, in contrast, is smooth, silky, and seductive, with bouncy rhythms backed by synthesized strings, in the manner, perhaps, of Kenny G. This dispute involves something more than the usual hip-hop polemics about "keeping it real" and not selling out, for the Infesticons' anti-jiggification crusade has little to do with mainstream hip-hop notions of "being real"; like other exponents of what Kodwo Eshun calls Black Atlantic Futurism, the Infesticons reject that "compulsory logic" according to which "the 'street' is considered the ground and guarantee of all reality" (4). Indeed, the very "gangsta" rappers who advertise their "love for the streets" (Dr. Dre) tend also to be the ones who boast the loudest about their jiggy lifestyles. This dilemma has deep roots in American black culture. Black people have it so hard in America, it's scarcely surprising how much emphasis is placed upon gettin' paid. The more precarious and hard-won the success, the more ostentatiously you will want to celebrate its rewards. In such conditions, jiggification is all but inevitable. The Infesticons see the bourgie lifestyle as a political betrayal, but they cannot deny its continuing pull. That's why their victory over the Majesticons is never final. As chief Infesticon Mike Ladd ruefully says at the very end of the album: "I guess I lost, 'cause I'm still mad at you."

Monadology. What does it really mean to be connected? What does it mean to be a node in the network? What does it mean to be present to others virtually, through a terminal? Many theorists have noted how presciently Leibniz, in his *Monadology*, anticipates the ontology of cyberspace. As Michael Heim rhapsodically expresses it, "monads may have no windows, but they do have terminals... The monad knows through the interface... Each monad represents the universe in concentrated form... Despite their ultimately solitary character, the monads belong to a single world" (96–98). Although he deplores the very situation that Heim celebrates, Slavoj Žižek describes it in much the same way: "Does our immersion into cyberspace not go hand in hand with our reduction to a Leibnizean monad which, although 'without windows' that would directly open up to external reality, mirrors in itself the entire universe? Are we not more and more monads with no direct windows onto reality, interacting alone with the PC screen, encountering only the virtual simulacra, and yet immersed more than ever in the global network, synchronously communicating with the entire globe?" (2001, 26). Heim and Žižek both point to the same Leibnizean paradox: we are simultaneously connected and alone. Indeed, our being each alone, rigidly separated from one another, is a necessary condition for our being able to log on to the same network. "Monads have no windows, by which anything could come in or go out," says Leibniz (179), but this very impermeability or opacity is what allows each monad to be "a perpetual living mirror of the universe" (186). The monad's closure is like the tain of a mirror,[4] the backing that allows it to reflect images—as a perfectly transparent window would not. Or, as Deleuze puts it, Leibnizean closure is "the determination of a being-for the world," in contrast to Heidegger's "being-in the world" (1993, 26). I do not find myself in the network, having fallen or been thrown. Rather, I exist for the network. I am predestined to it. From the moment I get connected, I am irreversibly bound to its protocols and its finality. Once that happens, it scarcely matters whether I am stuck, as Žižek fears, with "only the virtual simulacra," or whether, as Heim more optimistically maintains, I am now able to "get on line to reality" (98).

Are You Experienced? Perhaps the Experience Music Project, Paul Allen's "interactive music museum" designed

by Frank Gehry, will one day l
the organization of social space
much as Jeremy Bentham's Pan
the nineteenth. The building's ou
a colorful array of stainless ste
panels, arranged in broken lines
are supposed to evoke the shap
there is no correspondence betw
the actual insides of the buildin
solemnity. It is dominated by
"Sky Church," with its "85-foot
video frieze enhanced by sta
acoustics" (Experience Music P
hibition rooms branch off from
rock 'n' roll. When you first ent
high admission fee, you are
computer-cum-audio-device, c
and keypad. This is your person
pointing and clicking and ente
little lectures and sound sampl
catches your fancy. In this way,
and recodes music that was ori
electronics.[5] There is a sound bi
seum; you can make your wa
and in whatever order you cho
on display range from a guitar s
a costume worn by Queen Latif
early punk show to a Roland (a
these things is particularly int
combination of multiple items
milling crowds, and all the ske
ted over your headphones, lea
load. The Experience Music F
the space of an educational mu
reality entertainment. This spa

put us under surveillance,
ist for it, already. Deleuze
through the network as "a
ting moulding continually
the next, or like a sieve
int to another" (178–79). In
than tactile. It invites our
nes it is even touchy-feely.
oft's "Home of the Future."
y aspect of its environment
r eyes, turns on soft lights
s to commands. It teaches
pping and keeps an eye on
iature cameras" in every
always invoke "make pri-
anybody else to see you
itself has to know where
No more need for classified
s; discreetly and intimately,
hing.

t. The old disciplinary so-
dualistic. It was organized
prisons, of course, but also
arracks. But there is more.
Foucault says, "was also a
ermanent documentation"
nd regimented body, there
y, a *corpus* of meticulously
tion of these records is to
dies they refer to, by accu-
dge" about them and by
ic typology of delinquents"
provides a schema of norms
s upon which individuals

solitary character, the monads belong to a single world" (96–98). Although he deplores the very situation that Heim celebrates, Slavoj Žižek describes it in much the same way: "Does our immersion into cyberspace not go hand in hand with our reduction to a Leibnizean monad which, although 'without windows' that would directly open up to external reality, mirrors in itself the entire universe? Are we not more and more monads with no direct windows onto reality, interacting alone with the PC screen, encountering only the virtual simulacra, and yet immersed more than ever in the global network, synchronously communicating with the entire globe?" (2001, 26). Heim and Žižek both point to the same Leibnizean paradox: we are simultaneously connected and alone. Indeed, our being each alone, rigidly separated from one another, is a necessary condition for our being able to log on to the same network. "Monads have no windows, by which anything could come in or go out," says Leibniz (179), but this very impermeability or opacity is what allows each monad to be "a perpetual living mirror of the universe" (186). The monad's closure is like the tain of a mirror,[4] the backing that allows it to reflect images—as a perfectly transparent window would not. Or, as Deleuze puts it, Leibnizean closure is "the determination of a being-for the world," in contrast to Heidegger's "being-in the world" (1993, 26). I do not find myself in the network, having fallen or been thrown. Rather, I exist for the network. I am predestined to it. From the moment I get connected, I am irreversibly bound to its protocols and its finality. Once that happens, it scarcely matters whether I am stuck, as Žižek fears, with "only the virtual simulacra," or whether, as Heim more optimistically maintains, I am now able to "get on line to reality" (98).

Are You Experienced? Perhaps the Experience Music Project, Paul Allen's "interactive music museum" designed by Frank Gehry, will one day be seen to have prefigured the organization of social space in the twenty-first century, much as Jeremy Bentham's Panopticon anticipated that of the nineteenth. The building's outside is cheerily aggressive: a colorful array of stainless steel and painted aluminum panels, arranged in broken lines and swooping curves that are supposed to evoke the shape of an electric guitar. But there is no correspondence between this iconic exterior and the actual insides of the building. The interior reeks of faux solemnity. It is dominated by the pompous, overbearing "Sky Church," with its "85-foot-high ceiling and a massive video frieze enhanced by state-of-the-art lighting and acoustics" (Experience Music Project). Several floors of exhibition rooms branch off from this shrine to the glories of rock 'n' roll. When you first enter the museum and pay the high admission fee, you are given a heavy handheld-computer-cum-audio-device, complete with headphones and keypad. This is your personal guide to the exhibits. By pointing and clicking and entering numbers, you can get little lectures and sound samples relating to anything that catches your fancy. In this way, digital technology captures and recodes music that was originally produced by analog electronics.[5] There is a sound bite for every item in the museum; you can make your way through them selectively and in whatever order you choose. The fetishized objects on display range from a guitar smashed by Jimi Hendrix to a costume worn by Queen Latifah, and from a poster for an early punk show to a Roland (analog) synthesizer. None of these things is particularly interesting to look at, but the combination of multiple items, cramped exhibition cases, milling crowds, and all the sketchy information transmitted over your headphones, leads to a kind of sensory overload. The Experience Music Project seamlessly conflates the space of an educational museum with that of a virtual-reality entertainment. This space is simultaneously public and private. It is always filled with people waiting in line, craning their necks to look at one exhibit or another or milling about in rather cramped quarters. Yet these people never interact with one another, for each of them is lost in his or her own little private world, listening to individually selected sound commentaries. The space is united visually and outwardly, but it is fragmented aurally and inwardly. Just as in the *Monadology*, each person is isolated and alone. Yet they are all wired into the same network. The network can function in this manner, without contradiction, because each person is isolated and alone in precisely the same way.

From Surveillance to Control. Foucault memorably describes the Panopticon as an "instrument of permanent, exhaustive, omnipresent surveillance, capable of making all visible, as long as it could itself remain invisible...a faceless gaze that transformed the whole social body into a field of perception" (1979, 214). But that was then—the nineteenth century—and this is now. Deleuze says that Foucault "was actually one of the first to say that we're moving away from disciplinary societies, we've already left them behind. We're moving toward control societies that no longer operate by confining people but through continuous control and instant communication" (1995, 174). Once we have all been connected, there is no longer any need for the Panopticon's rigid, relentless, centralized gaze. The new forces of control are flexible, slack, and distributed. In a totally networked world, where every point communicates directly with every other point, power is no longer faceless and invisible. Instead, it works in plain sight. Its smiley face is always there to greet us. We are fully aware that its eyes are looking at us; it even encourages us to stare back. Rather than shrouding itself in obscurity and observing us in secret, the network offers us continual feedback even as it tracks us. It does not need to put us under surveillance, because we belong to it, we exist for it, already. Deleuze describes this process of control through the network as "a *modulation,* like a self-transmuting moulding continually changing from one moment to the next, or like a sieve whose mesh varies from one point to another" (178–79). In this way, control is less visual than tactile. It invites our hands-on participation. Sometimes it is even touchy-feely. Just look, for instance, at Microsoft's "Home of the Future." This "smart" house tailors every aspect of its environment just for you. It "looks into your eyes, turns on soft lights and mood music, and responds to commands. It teaches piano, helps do the grocery shopping and keeps an eye on the kids." The house has "miniature cameras" in every room, but not to worry: you can always invoke "make private" mode, if you don't want anybody else to see you (Heim 2000). Only the network itself has to know where you are and what you are doing. No more need for classified FBI files and secret police reports; discreetly and intimately, the network takes care of everything.

The Politics of Confinement. The old disciplinary society was representational and dualistic. It was organized around spaces of confinement: prisons, of course, but also schools, factories, and military barracks. But there is more. "The penitentiary Panopticon," Foucault says, "was also a system of individualizing and permanent documentation" (1979, 250). For each confined and regimented body, there is a corresponding textual body, a *corpus* of meticulously compiled documents. The function of these records is to characterize and identify the bodies they refer to, by accumulating "biographical knowledge" about them and by placing them within "a systematic typology of delinquents" (252–53). This typology, in turn, provides a schema of norms and deviations: a series of forms upon which individuals are induced to model themselves, thereby transforming themselves into subjects. In this way, the documents of disciplinary power are coextensive with all of social space. The disciplinary archives constitute a cohesive, totalizing representation of society and of every individual within it. Each person has an eerie double, in the form of a police file dedicated specifically to him or her. The disciplinary subject is thus what Foucault calls "a strange empirico-transcendental doublet" (1970, 318). It is both an object of scientific and forensic knowledge and a free, responsible agent. On one side, there are bodies, spaces, and empirical cause-and-effect relations. On the other side, there are bureaucratic forms, police dossiers, and transcendental conditions of possibility. These two sides correspond point by point. Yet they remain incommensurable. The map is not the territory. The dossier is not the prisoner. The data in your file can never establish your innocence but only determine your guilt. Indeed, as Kant and Kafka alike suggest, every subject is guilty a priori because its empirical strivings are never equal to the transcendental and unconditional demands of the Law.[6] The police derive their entire authority from this impasse.

The Politics of Performance. This all changes in the new control society. The network does not need any sort of a priori transcendental regulation. Instead, it is immanently self-organizing; that is to say, it operates according to "just-in-time" cybernetic feedback mechanisms. There may well be more surveillance than ever, but this surveillance no longer leads to an archive of "permanent documentation" that doubles actual existence. Instead, the results of surveillance are immediately fed back into the system. Surveillance records do not merely record past behavior, nor do they provide typological models to be applied to future behavior. Rather, the accumulated data works to manipulate behavior directly, in real time, in the immediate present. There is no longer any duality between data, on the one hand, and bodies to which those data would refer, on the other. Since I myself am already just a collection of data, it makes no sense to collect data about me. Surveillance records are no longer about anyone or anything. They are performative instead of constative; it is not what they say that matters, but what they do. It is precisely, and exclusively, under such conditions that we may say that everything in the world is *information*. The subject no longer exists as an "empirico-transcendental doublet"; its structure has collapsed back onto a single plane. All the familiar features of the network follow from this collapse into immanence: representation gives way to simulation, creation *ex nihilo* is displaced by mixing and sampling, murky depths give way to glittering surfaces, the service economy eclipses industrial production, and permanent employees are displaced in favor of "skilled New Economy workers day-trading their careers" (Conlin).

More Stately Mansions. Foucault argues, in the first volume of *The History of Sexuality*, that the bourgeoisie did not—as has often been assumed—invent sexual repression to control the behavior of the lower classes: "the deployment of sexuality was not established as a principle of limitation of the pleasures of others by what have traditionally been called the 'ruling classes.' Rather it appears to me that they first tried it on themselves . . . [The deployment of sexuality] has to be seen as the self-affirmation of one class rather than the enslavement of another: a defense, a protection, a strengthening, and an exaltation"; only subsequently did it trickle down to the lower classes "as a means of social control and political subjugation" (1978, 122–23). Much the same can be said about the current deployment of networking technology. Bill Gates is no Big Brother. He is more like Augustus Caesar or Saddam Hussein, erecting grandiose monuments to himself. Indeed, he went to great pains and enormous expense (something like $70 to $100 million) in order to equip his own home first of all with the very control devices that he hopes to sell, eventually, to the rest of us. Microsoft's "Home of the Future" is clearly modeled upon Gates's actual mansion: "as you move about the house, your choice of art appears on high-definition television monitors. Music, lighting, and climate settings all tag along, too. A small pin you wear lets the system know who and where you are. You can go to a computer terminal to pick out a movie or television program. It will follow you to the nearest screen. Only the phone nearest you will ring, assuming you've told the computer you're taking calls at all" (Folkers). Though video images follow you anywhere in the house, there is also a dedicated TV room, for family viewing; it "includes a video wall composed of 24 projectors and monitors. (Just in case Gates gets the urge to watch 24 movies at the same time)" (Drudge).[7] In this way, Gates's networking and control technology, like the sexual technology of the nineteenth-century bourgeoisie, is "a political ordering of life, not through an enslavement of others, but through an affirmation of self." And the point of such a technology, Foucault says, is always to produce "a body to be cared for, protected, cultivated, and preserved from the many dangers and contacts, to be isolated from others so it would retain its differential value" (1978, 123). So it's not that Bill Gates is scheming to oppress us; it's just that our oppression and subjugation is an unavoidable consequence of his own self-glorification.

Video on Demand. The video surveillance cameras are everywhere, keeping permanent watch over spaces both public and private. If the old disciplinary society of the nineteenth century was destined, as Mallarmé suggested, to end up in a book,[8] then the control society of the twenty-first is intended to be displayed on multiple video monitors. Instead of a single, self-contained Text that would comprehend the entire world within its pages, we face an infinitude of images, flickering on an infinitude of screens, each presenting its own singular point of view. Each video camera is a monad. Although every monad ultimately "represents the whole universe," Leibniz says, it does this only in a confused manner. It "represents more distinctly" just a small portion of the universe: its own "body," or immediate neighborhood (189). The network, therefore, has no Panopticon, no single privileged point of reference. The apparatus of surveillance is fragmented, multiplied, and widely distributed. Like God, or like infinite space in Pascal's frightful vision, its center is everywhere, and its circumference nowhere.[9] The old disciplinary surveillance system accumulated facts and figures. Its purpose was to see, and remember, everything that happened. But video surveillance in the age of the network goes further. It watches over even the emptiest expanses of space, and it registers vast stretches of time in which *nothing whatsoever happens*. The purpose of video surveillance is not to record crimes and other dangerous events so much as to prevent these events from happening in the first place. As J. G. Ballard puts it, "the cameras are there to deter criminals, not catch them" (2001, 58). If undesirable behavior is dissuaded right from the outset, it doesn't need to be disciplined later on. The result of this policy is a vast archive of video footage, most of which is devoid of incident, and most of which nobody ever looks at. But such images do not need to be seen. They are simply there, accumulating endlessly in vast archives, splendid in their emptiness and indifference.

Street Theater. Video surveillance seems to develop according to a principle of rarefaction, sort of like an inversion of Moore's Law. Video technology, like computer technology, is continually getting cheaper and smaller and more powerful. As a result, surveillance cameras proliferate with ever increasing speed. But these cameras do their job of dissuasion too well. The more cameras there are, the less crime there is for them to record. More and more video footage is gathered, but less and less ever happens on the tapes. What is to be done in the face of such a dearth of actual events? One response is offered by the Surveillance Camera Players (SCP).[10] This is a group of activists who "appreciate how boring it must be for law enforcement officers to watch the video images constantly being displayed on the closed-circuit television surveillance systems that perpetually monitor our behavior and appearance all over the city. The only time these officers have any fun watching these monitors is when something illegal is going on. But the crime rate is down... for untold numbers of police surveillants, there is less and less to watch—less and less to watch out for—every day" (1995). The SCP therefore propose "to present a specially-designed series of famous dramatic works of the modern period for the entertainment, amusement, and moral edification of the surveilling members of the law enforcement community" (1996). The works they have thus performed before surveillance cameras in public places generally involve meditations on totalitarianism and unbridled power. They include such classics of the theater as *Ubu Roi* and *Waiting for Godot*, as well as adaptations of texts ranging from Orwell's *1984* and *Animal Farm* to Wilhelm Reich's *The Mass Psychology of Fascism*. Of course, the SCP can never be sure whether or not they are being watched at any given moment. But even if their performances go unseen, they still address an implied audience. In this way, they conjure the audience out of thin air, compelling it to materialize. This audience is the hidden workforce behind the apparatus of surveillance. It consists,

among others, of "security personnel, police, school princi-
pals, residents of upper-class high security neighborhoods,
and the producers and salespeople of the security systems
themselves" (1995). The Surveillance Camera Players work
a kind of homeopathic magic. By giving the surveillance
apparatus something to look at, they contaminate and par-
asitize the network. They insinuate themselves within the
empty flow of images and give it a destination and a name.

No Privacy. "People have to wake up to the fact that
there isn't any anonymous usage of any communications
services," says one security analyst. "They have to get over
that" (Fox Market Wire). If everything is connected, nothing is exempt from surveillance. Software is already in place
to track your e-mail and catalog the pages you visit with
your Web browser. The FBI knows what you are doing,
and Microsoft and the media conglomerates aren't far behind. The scrutiny will only get more intense in the years
to come. As Julian Dibbell notes, the very techniques that
were initially developed to protect privacy are now being
used as weapons against it. For instance, steganography
(the art of concealing a message within other data, like an
image or sound file, so that third parties do not even know
that the message is being sent) is now being used in order
to create digital watermarks: "by weaving encrypted copyright information and serial numbers into the binary code
of photos, songs, and movies, rights owners can sear a sort of
virtual brand into their property... Some watermark schemes
haunting the research journals propose sending mark-
hunting Web spiders out to troll for content pirates and ID
them for prosecution." This development shouldn't come
as a surprise. Even in the heady days of the early 1990s, the
Cypherpunk movement, which championed digital encryption, was much more about secret bank accounts and anonymous financial transactions than it was about any other

sort of privacy or freedom. The eventual victory of advocates over the United States government did nothing but make the Internet safe for retail shopping.[11] Encryption and digital watermarking are indeed quite useful, if you want to protect monopolistic property rights. But that is not the same as privacy. The largest effect digital encryption has on personal security and freedom of expression is to restrict them. You can run, but you can't hide.

Transparency. If universal surveillance is unavoidable in the age of the network, then the next best thing might well be to make sure that the results of this surveillance are available to all. That is what David Brin proposes in his book *The Transparent Society*. It is futile to try to protect privacy by encrypting personal data or by otherwise placing limits on the flow of information, for the network tears down all barriers in its path. Efforts to restrict the flow of data in the name of privacy will only accord greater impunity to the most powerful, while leaving everybody else still vulnerable to intrusion. Brin therefore proposes a policy of "reciprocal transparency" (81) and "mutual accountability" (322). If footage from surveillance cameras were openly available to everyone, the cops would still be watching every street corner, but we the people could also keep an eye on what the cops were doing. If all financial transactions were publicly posted, then no one could divert my money without my knowledge; protection against fraud would actually be stronger than it is with today's password-protection schemes. Brin's proposal is of course anathema to the Cypherpunks, who dream of never having to pay taxes or deal with other people face-to-face. But it is equally repugnant to the Surveillance Camera Players, for whom it amounts to "the total surveillance of all by all ... the mass murder of social life" (1999). For the one group, Brin's proposal would destroy individual autonomy; for the other, it would destroy community. A better criticism might be that Brin fails to adequately consider the role of the marketplace. Under his plan, information would not be "free"; it would be a commodity, available if and only if you had the money to pay for it. In fairness, Brin is not unaware of this difficulty; he tries hard to balance the claims of ownership and access. On the one hand, he insists that the "easy *availability* of information" does not mean "never having to pay for what others worked to produce" (103). He hopes that businesses will be able to set up an equitable system of micropayments so that the price of information will not be exorbitant. On the other hand, given how easy it is to make digital copies and how difficult it may well be to establish such a payment system, he always prefers to "err in favor of openness" (105). Brin himself is eminently reasonable; the problem is that money and information are not. They tend to proliferate virally, deliriously, beyond all measure. Money and information transform whatever they encounter into more money and more information. In Brin's "transparent society," money and information themselves are the sole things that remain mysterious and opaque.

Irrational Exuberance. Manuel Castells argues that, in the new network economy, "financial markets, by and large, are outside anyone's control. They have become a sort of automaton, with sudden movements that do not follow a strict economic logic, but a logic of chaotic complexity" (2001, 87). Sober empirical research leads Castells to conclusions matching those deliriously divined by Baudrillard: that the market no longer measures economic value, for "it bears no relation to any reality whatever" (2001, 173). Inverting Marx, Baudrillard insists that exchange value no longer "represents" use value; instead, use value is merely a chimera, whimsically "produced by the play of exchange value" (103). It is very much in this spirit, though without

Baudrillard's dubious metaphysics, that Castells argues against the common belief that the dot-com stock bubble of the late 1990s was based upon absurd overvaluations of companies that had no profit potential. "I think the 'bubble' metaphor is misleading," Castells writes, "because it refers to an implicit notion of natural market equilibrium, which seems to be superseded in the world of global financial markets operating at high speed, and processing complex information turbulences in real time" (105–6). Financial markets never reach a point of equilibrium; indeed, there is no longer any way to define such a point, even in principle, for the logic of financial markets today is entirely phantasmatic and nonrepresentational. Stocks do not represent values that exist in the real world; rather, financial speculation itself is what generates all those real-world values in the first place. "Investment is led by the growth of value of stocks," Castells says, and "not by earnings and profits" (85). This is why traditional measurements of the proper value of stocks, like price/earnings ratios, are no longer relevant in the new global economy. It follows that the crash that started in March 2000 should not be regarded as a "correction" or a return to reason, for the bear market was no more grounded in objective notions of value than was the extravagant bull market that preceded it. All in all, Castells writes, "the lesson does not seem to be one of irrational exuberance followed by sudden temperance, but, on the contrary, of jittery behavior structurally determined by globalization, deregulation, and electronic trading" (89). Financial markets are "systematically volatile" (90); they are systematically resistant to any form of stabilizing systematization. It may be precisely for this reason that "the market" is so often substantialized in business news and other financial discourse, as if it were an agent with moods and volitions of its own ("the market was jittery today..."). Referring only to itself, and yet affecting everything and everyone, and leaving trains of devastation in its wake, the stock market is as capricious and unpredictable as only another person could be: somebody who fascinates and captivates me, but whom I do not really know all that well.[12]

Zeroes and Ones. Today it seems naive to believe, as many people did in the early days of the Internet, that "information wants to be free." Now we know better: as the space traders in Ken MacLeod's science fiction novel *Cosmonaut Keep* like to say, "information wants to get paid" (39, 129). The dilemma of information management is a binary one: how to accumulate information, on the one hand, and how to limit access to it, on the other. Information is not like those old-fashioned commodities that get cheaper as more of them are produced. Rather, the greater the amount of information that is gathered, the more correlations and cross-references it potentially contains; and consequently, the more each new bit of information is worth. Value results from abundance, not from scarcity. That is why we are under such relentless pressure to digitize more and more, until everything is "transcoded"[13] into binary data, and made available for unlimited replication and manipulation. At first, this might seem like a DJ's ultimate dream: the whole world is there to be sampled and remixed. But as hip-hop producers unhappily learned in the late 1980s and early 1990s, even the smallest digital sample can be traced, and every sample therefore must be licensed and paid for. That is why no music released today has the sonic density and richness of De La Soul's *3 Feet High And Rising* (1989). For this was the album that "broke ground in the legal arena ... After the Turtles won a lawsuit against De La Soul for the use of [just four notes from] their song 'You Showed Me,' all samples had to be cleared before a song could be released."[14] A 1992 court ruling against Biz Markie further solidified this requirement (McLeod 88). Hip-hop production has never recovered from this ruling. Music, unlike traditional publishing, is now subject to prior restraint. Evidently, the rights of free speech and fair use do not apply in the digital domain. Such stringent restrictions might seem to contradict the extravagant promise of digital technology. But in fact, they follow quite logically from it. For the massive accumulation of digital data is also an act of appropriation. Turning the heterogeneous contents of the world into zeroes and ones, uniform bits of information, is a way of claiming ownership of those contents. Digitization goes hand in hand with privatization. It's our version of what Marx called primitive accumulation (1992, 873–95). Just as British landlords, at the start of the sixteenth century, expropriated the peasants and enclosed formerly common lands, so multinational corporations, at the start of the twenty-first, are appropriating data that used to be in the public domain and turning culture itself into a private preserve. They are compiling records about every aspect of your life: what Web sites you visit, what items you buy, what music you listen to, how many children you have. They are converting the very fabric of everyday life into a series of protected trademarks. And they are patenting and marketing the genes of plants and wildlife whose beneficial properties are known to indigenous peoples— and sometimes even the genes of the indigenous peoples themselves.

Codes for the Cataclysm. "I wanna devise a virus / To bring dire straits to your environment / Crush your corporations with a mild touch / Trash your whole computer system and revert you to papyrus." These words are from the hip-hop album *Deltron 3030*, a collaboration between rapper/MC Del tha Funkee Homosapien, producer Dan the Automator, and DJ Kid Koala. It's a science fiction concept album, set in the year 3030 AD. The apocalypse has already happened, but "the devastation wrought by WW4" doesn't seem to have interfered with business as usual. Spaceships are as common in the year 3030 as automobiles were in the year 2000. Robots have overrun every corner of the galaxy. And consumers face a continual barrage of aggressive multimedia advertisements, proposing ever greater degrees of cyborg hybridization: "upgrade your gray matter / 'Cause one day it may matter." But if human mental capacities are pushed to the limit in this high-tech world, it is not because of competition from advanced artificial intelligences, nor even because of information overload. What really makes a difference is the distribution of information rather than its mere quantity: power consists in controlling access to the best databases, and extracting payment for their use. Just as in Burroughs's Ba'dan and in *Transmetropolitan*'s City, the forces of imposed scarcity continue to rule, even amidst material and informational plenitude. New technologies have not diminished the need for struggle, even though they have radically changed the shape of the field of battle. Conditions in the society of 3030 are still "much like America of today, where everyone is stuck under the thumb of 'the man,'" Mega-corporations monopolize the flow of information. Even the news, we are told, is "a wholly owned subsidiary of Microsoft Inc." A "world-wide government" maintains order through repression and violence; "turbulence and murder" has become "an everyday occurrence." And "global apartheid" is the prime directive of the new galactic order.

Rhythm Warfare. What can be done to oppose this universal terror? In the world of *Deltron 3030*, "there is no 'rebel alliance' or any such nonsense as one might imagine if they prescribe [sic] to the Star Wars theory of the apocalypse." Resistance can only be indirect, disrupting the system "with a mild touch." It comes in the form of verbal and sonic guerrilla warfare: Del's tongue-twisting rhymes and language-contorting raps, Koala's offbeat scratches, and the Automator's slow beats and eclectic samples. Hip-hop is an art of recombination; its materials are not words and sounds newly made, but already-existing fragments of commodity culture, wrenched violently out of their previous contexts. Or as Walter Benjamin puts it, "blasted out of the continuum of history" (1969, 261). Del and his crew are not Luddites; after all, it takes considerable technical expertise to write a virus powerful enough to "trash your whole computer system." They take technology, therefore, and turn it back against itself. They scramble linguistic and computer codes to create a soundscape of malice and foreboding. They unwind the skeins of digital control, feeding code back onto itself until it *rebecomes analog*. For, as Brian Massumi insists, the digital is never autonomous or complete unto itself; it is always "sandwiched between an analog disappearance into code...and an analog appearance out of code" (138). Del's computer virus is designed to provoke just such a rebecoming or reappearance. Hip-hop is a kind of linguistic and cultural hacking. Its experiments are, at one and the same time, interventions in digital code, and analog movements in and through physical space. And so, Del's "adventures in sonic fiction" (Eshun) are also cosmic voyages, as his rogue spaceship flits across the galaxy, in a fugue of demented loops and complex evasion routines. We are far from the all-embracing warmth of George Clinton's Mothership, let alone the regal majesty of Sun Ra's intergalactic vessel, powered by his harmonic progressions.[15] Del's ramshackle craft is both more and less than these. It lacks the alien splendor and power of Clinton's and Ra's vehicles from beyond; but perhaps it is more adept at maneuvering through the spaceways of tomorrow, filled as they are with private preserves and no-trespassing zones, random bits of detritus and spam, and intimidating police cruisers.

Not with a Bang, but a Whimper. Outright censorship is an old-fashioned way of controlling access to information. It's a holdover from the age of print. It was effective then; the authorities could easily seize a printing press and burn all extant copies of a newspaper or a book. But such an approach doesn't work very well in the age of the network. It goes against the grain of digital technology. On the Net, it is far too easy to make copies of offending material and distribute the copies to servers everywhere. A subtler method of information control is therefore needed, one that goes along with digital technology, instead of making futile attempts to block it. That is why the future lies with Digital Rights Management: encryption and registration schemes, and proprietary standards. Information itself is the best tool for regulating information. "The digital language of control," observes Deleuze, "is made up of codes indicating whether access to some information should be allowed or denied" (1995, 180). These codes are active at every moment. The point is not to stop the flow of information, but rather to channel and direct it, and even positively encourage it, by embedding surveillance and control mechanisms within it. Every bit of information can be tagged and licensed. Every transaction it passes through can be monitored on the network. Every attempt to access the information can be managed, generally on a pay-per-use basis. Controls can be built directly into the hardware and the software in a manner invisible to the consumer. Under a recent Microsoft plan, for instance, "your PC will just go ahead and check you're licensed to play whatever music (or use whatever data) you acquire, and will only need to bother you whenever it needs a credit card number" (Lettice). Jaron Lanier similarly worries that the next generation of audio and video equipment, as well as computers, may well be designed to "play only music authorized to be heard at a given time and place . . . unregistered material simply [won't] play on any video screen or speaker in the land."

From Statute to Contract. The American entertainment industry is well on its way to obtaining a more thorough control over speech and expression than Stalin or Ceausescu ever dreamed of. It is happening now, in three easy steps. First, all speech and expression is translated into digital bits, codified in the form of information. Once this is accomplished, speech and expression have no special privileges; information is just another commodity. Digital code is a universal medium of exchange, like money: it makes any given object commensurate with any other. The ideal of modernist aesthetics is thus ironically realized: in the digital realm, form and content are one. It is no longer possible to make the old distinction between ideas (which cannot be copyrighted) and specific expressions of those ideas (which can). Everything is code, or specific expression. In the second place, all this information is privatized, together with the network through which it circulates. There are limits to what a government can do, but these limits do not apply to private corporations. Ownership is its own authority. In the United States, for instance, the First Amendment to the Constitution guarantees your right to pass out political leaflets on a public street corner, but not in a privately owned shopping mall. Burning the American flag is protected speech, displaying a defaced Barbie doll or an image of Mickey Mouse with genitalia is not. In the third place, since information is always somebody's property, its use is governed by contract, instead of by statute. You have no intrinsic rights when it comes to information, but only temporary access, paid for in commercial transactions. Just as writers, musicians, and artists are merely "content providers," or workers for hire, so readers and listeners are just "end users," or licensees. If you don't like the terms of the licenses, you can always opt out of the contract; nobody is forcing you, after all, to listen to music, or keep up with the news, or look for a better job, or talk to people on the phone.

Oxygen. Information is like the air we breathe. It is the element we live in. It surrounds us on all sides, and we couldn't survive without it. Does this mean that information ought to be free, or at least a public good? A libertarian, free-market economist warns us against such logic. Steven E. Landsburg tells us of the many problems we face simply because the air is free. That is why we suffer from air pollution, for example. If the atmosphere were privately owned and sold on the open market, then we would be compelled to use it more efficiently. There would be a "powerful disincentive" to waste and pollution, Landsburg says, if only we would "have to pay for the right to breathe freely" (81). Landsburg modestly fails to note another advantage of his plan: it would also solve the problem of overpopulation. If too many people were competing for limited natural resources, then scarcity would cause the price of air to rise. And this, in turn, would reduce the consumption of air, especially by all those impoverished and inefficient breathers. In short order, the market would work its magic, and the population of the world would be cut back to a suitable equilibrium level.

Your Money or Your Life. When it comes to controlling and restricting speech, Singapore is far ahead of the rest of the world. It's the one place where this task has already been privatized, in accordance with free market imperatives. Opponents of the regime are not threatened with jail; they have far more serious things to worry about. If you criticize the Prime Minister, as Tang Liang Hong did in 1997 while running for a seat in Parliament, then you are likely to be sued for libel. You will not face criminal charges and imprisonment, but rather massive civil damages in a defamation suit (Safire). It's not the State that comes after you, but a private individual (like the Prime Minister himself) or a corporate entity (like his political party). Indeed, in Singapore "the ruling Peoples Action Party regularly uses the defamation laws to bankrupt those who dare to criticize the government" (Divjak). The advantages to this system are obvious. For one thing, criminal penalties like fines and jail terms are strictly limited by statute, but civil awards resulting from litigation are potentially limitless. Tang, for instance, was ordered by the court to pay the Prime Minister and his party a total of US$5.8 million. For another thing, while criminal law puts the burden of proof on the State, in a civil suit the standards of proof are much more flexible. This sometimes gives the proceedings an *Alice in Wonderland* quality. It was the Prime Minister who called Tang a chauvinist and a bigot; Tang was sued for slander because he had the impudence to deny these charges, thereby implying that the Prime Minister was a liar. Finally, litigation has the advantage that restitution is paid, not to the State, but directly to the aggrieved party. If Tang had actually had as much as US$5.8 million to lose, this money would have gone, not to government coffers, but right into the pocket of the Prime Minister himself. Singapore has pioneered a new conception of law for the privatized information age; can the United States of America, with its fanatical devotion to private enterprise, be far behind?

Corporate Free Speech. As information is accumulated and privatized, free speech increasingly becomes a matter of property law. As an individual, I am permitted to say less and less, because so much of what I want to say involves reference to material that is protected by someone else's copyright. The frequent citations that appear in this book are covered under the doctrine of "fair use."[16] But in the United States, the Digital Millennium Copyright Act (DMCA) of 1998 radically limits the scope of fair use, at least with regard to digital media.[17] I am subject to felony prosecution, for instance, not only if I have an MP3 file on my hard drive that is derived from a copyrighted source—which I do—but even if I have software that would make it theoretically possible for me to create such a file at some point in the future. The Consumer Broadband and Digital Television Promotion Act, before Congress as I write these lines in 2002, goes even further, making it a felony to manufacture, sell, or give away any electronic device that does not have a certified copy protection mechanism already built in. It's already a crime to imagine remaking a film with different actors (Litman 22, 32, 71, 75). Conversely, the drive to define the content of speech as private property has increased the rights and powers of corporations. United States courts recently struck down a law that would have prevented "credit bureaus, private investigators and information brokers" from selling data about prospective customers to retailers and advertisers without the customers' permission. This was considered a violation of the credit bureaus' First Amendment rights to free speech (Sanders). That is to say, if a company has my social security number, my driver's license number, and my credit record in its database, then this information is the company's property, and they are free to sell it to anyone, or otherwise do with it as they wish. Laws that try to regulate the concentration of media ownership have also been struck down as violations of the First Amendment. If companies like Comsat and AOL Time Warner own the broadband networks, then it is a violation of their free speech rights to compel them to carry any content on these networks that they would prefer to censor or omit. Corporate free speech is thus rapidly "becoming synonymous with the right of one monolithic media corporation to dominate any given market'" (Kumar). Similar arguments have also been made with regard to campaign finance reform. It would be a violation of the First Amendment to restrict the flow of political campaign contributions in any manner whatsoever. Spending one's own money is thus enshrined as the purest form of expression. And any speech not backed up by money is precisely worthless. This is Steven E. Landsburg's free-market utopia put into actual practice in the here and now.

The Real Meaning of Intellectual Property (I). In 1969, Columbia Records adopted a new advertising tagline: "The Man Can't Bust Our Music." I was fifteen at the time and pretty indifferent. But several friends, slightly older than I am, have independently told me that this was the moment when they first realized that anything whatsoever could be co-opted.

The Real Meaning of Intellectual Property (II). When Beyonce Knowles, the lead singer of Destiny's Child, was asked (during a TV interview in the summer of 2001) where she saw the group heading in the next five years, she replied that she hoped to be able to do more product endorsements.

Sic Semper Tyrannis. In the pages of Noir, K. W. Jeter earnestly advocates the death penalty for violators of copyright. He presents this policy as a necessary one, in our networked world where information is the major source of value, and people make their living by producing "intellectual property," rather than by manufacturing physical objects. "*Ideas* and/or *design* and/or *content*—whatever word, name, label one wants to use—if that [is] the most important thing in the world, that which determine[s] whether you eat or starve," then it needs to be defended like life itself (251). Downloading MP3s via Napster, or getting a science fiction short story from Usenet,[18] is pretty much the same thing as murder and deserves to be punished accordingly: "to steal [intellectual property] from someone [is] not to take some expendable, frivolous trinket off them, a van Gogh off the dining-room wall of some bloated plutocrat; but rather to lift some hard-pressed hustling sonuvabitch's means of survival, the only way he ha[s] of turning the contents of his head into the filling of his stomach" (254). Frivolous trinkets and van Gogh canvases might seem to make strange bedfellows. But they are both singular art objects that cannot be replicated. Their *aura*—defined by Walter Benjamin as the artwork's "presence in time and space, its unique existence at the place where it happens to be" (1969, 220)—exempts them from the usual circuits of economic exchange. On the other hand, Benjamin says, the aura "withers in the age of mechanical reproduction" (221). This withering is all the more pronounced in our current age of electronic, digital replication. Information, or intellectual property, has no aura; it is fully subjected to the rules of commerce and realized exclusively in the form of exchange value. The problem is that the same technology that makes information the predominant source of value also cheapens this very information by making it easy for us to replicate it at almost no marginal cost. This is an intensified version, for the information age, of what Marx (1993a) called "the tendential fall of the rate of profit." It leads to the dilemma described by John Perry Barlow (1994): "if our property can be infinitely reproduced and instantaneously distributed all over the planet without cost, without our knowledge, without its even leaving our possession, how can we protect it?" Barlow answers this question by concluding that the notion of copyright is obsolete. He proposes that intellectual property law be junked entirely and that digital content be given away for free. And he suggests that content providers or producers can best earn their livings by providing secondary services connected with their creative work: things like personal access, live performances, and frequent updates. Indeed, this is what academics already do today: I earn my living through my teaching and lecturing, and the main economic benefit that I get from my writing is that it may make me more in demand as a teacher and lecturer. Jeter, however, will have none of this. A professional writer who depends upon the actual sale of his books, he is not ready to give away the fruits of his primary labor. So he ridicules Barlow's proposals as "weird 'net-twit theorizing ... [a] half-baked amalgam of late-Sixties Summer of Love and Handouts, Diggerish free food in the Panhandle of Golden Gate Park, and Stalinist collectivization, lining up the kulaks and shooting them 'cause they're in the way of the new world order" (251–52). Jeter concedes Barlow's point about the replicating power of the new digital media. There may well be no good way to protect intellectual property on the level of software. But software is not the only means of protection. As Jeter points out, "there's a hardware solution to intellectual-property theft. It's called a .357 magnum. No better way for taking pirates off-line. Permanently. Properly applied to the head of any copyright-infringing little bastard, this works" (255).

Better Off Dead. In a debate during the United States presidential campaign of 2000, George W. Bush rhapsodized over the virtues of capital punishment. Speaking of some convicted killers, he said, with a self-satisfied smirk: "guess what's going to happen to them? They're going to be put to death. A jury found them guilty and it's going to be hard to punish them any worse after they get put to death. And it's the right cost; it's the right decision." Of course, this fatuous assertion just shows Bush's difficulty with what his father used to call "the vision thing." If Bush *fils* weren't so woefully unimaginative, it wouldn't be so difficult for him to figure out a way to "punish them . . . worse." Jeter points out that "the main problem with death as a negative motivational factor [is] that it [is] over too quickly, and not always painfully enough" (257). What's needed to make capital punishment really work as a deterrent is "to stretch death out in time, take it from a point to an extended process. And pump up the pain and humiliation factors" (257). In short, there must be a *"catastrophic price"* (266) to pay for copyright infringement. This is accomplished in the world of *Noir* by condemning information pirates to a lingering life-in-death. They are never granted the certitude and finality of actually being dead. Instead, they must face what Maurice Blanchot calls the interminability of dying: an "abyss of the present, time without a present," in which "they do not cease, and they never finish dying" (155). In Jeter's vision, the offender is immobilized but compelled to remain conscious. His limbs are lopped off, his torso is sliced open, his internal organs are removed—all without benefit of anesthetics. His brain and spinal cord are then extracted from the rest of his body, which is discarded as garbage. Just enough neural tissue is preserved to make sure that the offender retains his "basic personality structure and an ongoing situational awareness" (262), and above all his ability to feel pain: "the brain matter, the still-living remnant of the various pirates, was there for one purpose. To suffer" (261). Finally, the offender is transformed into a "trophy": the mass of neurons and synapses—all that remains of him—is encased within a speaker wire, a vacuum cleaner, a toaster, or some other "common household appliance" (264–65).

This "trophy" is handed over to the artist whose copyright the offender has violated. In this condition, the offender faces a virtual eternity of pain: "the essential brain tissue, and the consciousness and personality locked inside the soft wiring, could last decades, perhaps centuries" (266). And the artist is gratifyingly reminded of the violator's agony every time he or she listens to the stereo or makes a slice of toast. What's remarkable about this section of *Noir* is not just the amazing conception of the punishment mechanism, a sadistic machine to rival the one described in Kafka's "In the Penal Colony." Even more startling is the gusto and gleeful exuberance of Jeter's prose as he describes the torture process in loving, intricate detail, savoring it on page after page. Indeed, this is the only note of excess in what is otherwise a carefully balanced and perfectly plotted novel. It's also the only place in this bleakly dystopian book that Jeter allows himself any signs of hope, satisfaction, or pleasure.

Pirate Utopia. Programs like Napster—and, to a lesser extent, its various clones and successors—shook up the music industry at the end of the twentieth century. These programs permitted the easy, and nearly unlimited, copying and sharing of MP3 music files. For many of us, it was a sort of heaven: a sojourn in the paradise of music, a carnival of sonic promiscuity and miscegenation. "If ever there was a genuine pirate utopia online," Julian Dibbell writes, "it was Napster." For me, for Dibbell, and for many of Napster's fifty-eight million users, the program's heyday, from late 1999 to early 2001, was a time of unprecedented freedom and joy. People put their entire music libraries online, and downloaded hundreds, or even thousands, of additional tracks for free. I myself was only a casual, occasional user of Napster, but I still managed to add over 500 songs to my hard drive, or more than a gigabyte's worth of MP3 files. (And this figure doesn't even include the numerous files I listened to only once or twice, and then erased.) The music ranged from contemporary Top 40 pop hits (like songs by TLC and Destiny's Child), to vintage Detroit techno (like Underground Resistance) and the latest British dance tracks (by producers like Wookie and the Artful Dodger), to miscellaneous oldies (like psychedelic schlock from the late 1960s or the sexy 70s songs of Millie Jackson), to anonymous, totally unauthorized, and often brilliantly satirical remixes (like a file of Eminem rapping to the music of Enya). Once, as a test, I found and downloaded the entire contents of a 2-CD collection of Tupac Shakur's greatest hits in an hour and a half—less time than it would have taken to listen to it all. But I also found, just as easily, twenty-two-year-old, nearly forgotten New York punk hits by the likes of Richard Hell, Lydia Lunch, and James Chance. The sheer variety of stuff on Napster was amazing. Late at night, when everyone else in my neighborhood was asleep, I'd be tracking down some half-forgotten ditty from my youth or stumbling across some obscure, out-of-print electronic track that I had only vaguely heard of and that turned out to be shockingly innovative and fresh. Along the way, I'd meet, and exchange messages with, other collectors, from Germany or Australia, who shared my passion for Captain Beefheart, or Funky Four Plus One, or Cha Cha, or Os Mutantes, or Nick Drake. No doubt about it, this was cultural revolution. It seemed as if fifty years' worth of popular music—music of all styles, countries, genres, and periods—were now present, all at once, in the same virtual space, and free for the taking. In such a climate, I came to believe that sharing files with strangers was a virtuous act; even more, that circumventing copyright law was a high moral imperative. All this music was culture itself, the very element in which we lived and moved; forcing us to pay for its ubiquitous manifestations would be as odious as forcing us to pay for the air we breathe, in the manner that Steven E. Landsburg recommends.

The Bottom Line. My experience with Napster is why I am forced to admit, however reluctantly, that Jeter has a point in proposing the draconian measures that he does against copyright violators. Anything less ruthless and violent is unlikely to work. Think about it for a moment. All of the files that I downloaded via Napster, without exception, had at least this one thing in common: they all consisted of copyrighted material, for which the permission to reproduce had never been granted. I estimate that it would have cost me US$750 at the very least, and probably more like US$2500, to obtain all that music legally in the form of CDs. Multiply this by fifty-eight million users, and the figure comes out to nearly US$150 billion per year. You can see why Napster put the recording industry into a panic. But I couldn't care less. Like most other Napster users, I didn't feel the slightest compunction about what I was doing. I totally rejected the industry's contention that downloading music was anything like stealing actual CDs from a record store. For one thing, I never took any physical objects. I always paid full price for my Internet connection, as well as for the music's physical media: the computer equipment I used, the hard drive onto which I downloaded the files, and the CD-R discs onto which I subsequently burned them. For another thing, it wasn't just a matter of saving money: I positively exulted in the fact that I was sticking it to Sony and Universal—and for that matter, to Destiny's Child and Eminem as well. Telling off the entire entertainment industry in this way was (as they say in those MasterCard commercials) priceless. When attitudes like mine are so widespread, it will take a lot of forceful persuasion to alter them. It's a situation where only the most extreme positions, on either side, are viable. As Jaron Lanier puts it, "if we make

Napster-like free file sharing illegal, we'll have to rid ourselves of either computers or democracy. You can't have
both." And you know which alternative copyright holders
will choose. Nothing less will have any effect. That is the
logic behind Jeter's proposal. I admit it: if I had had reason
to fear that the fate of "trophy-ization" might have been
awaiting me for pirating copyrighted material, then I almost
certainly wouldn't have done it.

A General Theory of Evil. Jeter is closely attentive to
the psychology of copyright piracy. He wants to know what
it is that drives people like me to traffic in illicit intellectual
property. Ultimately, Jeter proposes "a general theory of evil"
(239), with copyright violators at the apex. They are far more
depraved and unnatural than contract killers, child moles-
ters, or even corporate executives like Harrisch of Dyna-
Zauber. Death may be an effective deterrent for kidnappers
and murderers, but it is actually more of an incentive for
the average software pirate or Napster user. For in Jeter's
view, copyright violation is motivated largely by resent-
ment: "there were certain people who loved the art—the
music, the books, the pictures, whatever it might be—but
who actively hated the creators of the same. Hated them
from envy, jealousy, spite—from just that gnawing, infuri-
ating sense that the creators could do something they
couldn't" (239). As Jeter pursues his analysis, it becomes
ever more apparent that this is the basic condition of aesthetic reception. To love a work of art is to be consumed
with rage against that work's creator. Relations among
artists themselves often work this way, if we are to believe
Harold Bloom's theory of "the anxiety of influence." An
artist claims a place for him or herself, Bloom says, by reac-
tively and spitefully "misreading" the work of his or her
precursors. Jeter extends this analysis to what Bloom
would call the "weak misreadings" of ordinary readers, lis-

teners, and viewers. Such people "don't steal things," Jeter says, "just so they can have them" (240). Indeed, they will even "shell out nearly the same amount or even more to a pirate, some copyright rip-off specialist, rather than see the same money or even less go to the rightful creator" (239). The actual purpose of copyright infringement is less to get something for nothing than it is to assert "that the books and the music and the paintings and everything else really belonged to the thieves, that it was all theirs by right; in some strange way, the thieves and not the creators had brought it all into being" (240).

The Pleasure of the Text. Every copyright violator would thus be able to say, along with Roland Barthes, that "the birth of the reader must be requited by the death of the Author" (55). When readers, listeners, and viewers download works for free, they are just taking the logic of Barthes's aesthetics to its fatal conclusion. A work does not "belong" to its creator, Barthes argues, because it is "a multi-dimensional space in which are married and contested several writings, none of which is original: the text is a fabric of quotations, resulting from a thousand sources of culture" (53). That is to say, *all* texts are already doing implicitly what postmodern works (from hip-hop tracks to novels by Kathy Acker) do explicitly and overtly: they sample, they appropriate, they hybridize, they distort, they remix and recombine, the already-existing detritus of culture. For Barthes, this is a joyous and wholly affirmative process, one that he describes in rapturously sexual terms. But one man's *jouissance* is another man's poison. It is precisely this Barthesian, oedipal usurpation of the author's role that makes copyright violation a more despicable act, in Jeter's eyes, than rape, child molestation, or murder. For these latter crimes are not altogether devoid of creativity; they may well be driven by some terrible, twisted form of sexual desire. But copyright piracy cannot possibly be creative; it's a pure expression of negativity and hatred, of what Freud calls the death instinct, and Nietzsche, *ressentiment*. In the last analysis, Jeter suggests, the love of culture is really a concealed death wish. Copyright violators are ultimately motivated by "some sort of self-destruction imperative, sure and certain suicide by means of the law's swift, implacable enforcers" (247). We are attracted to works of art like moths to a flame. What we really want, when we think that we love a work of art, is for it to overwhelm us, trample us, and crush us into bits. We hate and resent creators, above all, because they see right through us: they understand our secret lust for annihilation, and they offer to fulfill it. That is why the death penalty is more a lure than a deterrent when it comes to copyright piracy. To put an end to copyright infringement once and for all, something harsher and crueler is needed: something "beyond death, beyond the notions of desirability for even the terminally self-destructive" (266). And hence, Jeter says, *"the creation of the trophy system"* (257). For the sake of his very survival as an author, Jeter is compelled to wage war against his own readers. In this vision of morality and aesthetics, Kant's Second Critique comes together with his Third. The implacable severity of the categorical imperative joins with the shattering intensity of the sublime. The authentic realization of the work of art can only be achieved by the suppression of sampling, through the full and rigorous enforcement of copyright law.

The Gift-Based Economy. "'Information wants to be free, huh?' McNihil didn't wait for an answer. 'Well, here's some info you can have for nothing.' He swung his fist in a hard, flat arc, landing it straight to the kid's nose, which exploded in a bright flower of blood" (196). This scene from *Noir* epitomizes Jeter's theory of value. Giving something away for nothing is the one unforgivable sin. Full and immediate payment for value received is the very basis of human sociality. Open-source free software and texts in the public domain are obscene. "The 'gift-based economy' had been a hippie dream, nice for exchanging information of no value, worthless itself for selling and buying anything *worth* buying and selling" (252). The question is precisely one of *worth*. In the globalized network society, the "free market" is the sole arbiter of value; there is no standard of worth, other than the economic one. If the market is "free," then nothing else can be. Everything has its price, as determined by the iron laws of supply and demand: "the laws of economics [are] as immutable as those of physics" (252). Jeter scornfully rejects anything that might smack of what Georges Bataille (1988) calls *unproductive expenditure*. On the one hand, Jeter says that true generosity and gift-giving are, lamentably, impossible. In a world of Darwinian struggle, "there are no favors. Nobody does favors for other people" (268). Behind each seemingly unselfish act, there is always an ulterior profit motive at work. When I receive a gift, I can be sure that the giver will make me pay for it, in one way or another. This truth may be ugly, but it is better to know it than to be played for a sucker. On the other hand, and at the same time, Jeter maintains that generosity and gift exchanging are positively evil. Gratuitous pleasure violates the harsh Calvinist ethic that gave birth to capitalism in the first place. To enjoy something without paying the proper price for it, without even wanting to pay such a price, is for Jeter the most loathsome form of depravity: "How could you be into something, into it enough that you wanted all you could get of it, and not want to pay for it?" (239). This is why the only thing that enrages Jeter more than people who violate copyright for personal gain is people who violate copyright just for the hell of it, without even making money from the process. The penalty that such people must pay for turning "intellectual property" into a gift is to be transformed into gifts themselves, in the form of trophies: this is the only form of gratuitous exchange that Jeter allows into his system. Any deconstructionist worth his or her salt will recognize the deliriously self-contradictory logic at work here: Jeter imagines a fantastic police apparatus just to repress a "gift-based economy" that should not even be possible in the first place. Such a vision is, as Derrida says in a similar context, "contradictorily coherent. And as always, coherence in contradiction expresses the force of a desire" (279).

Immutable Laws. Jeter's copyright militancy has at least this virtue: it leads him to point out, *contra* John Perry Barlow, that the question of intellectual property is not merely a technological one. It is political and economic, first of all. Barlow argues that information must be treated differently from real estate, simply because the latter is tangible and physical, while the former, ostensibly, is not. Jeter, more acutely, compares "intellectual property digitized on the wire" with "cattle [roaming] across grazing land" on the old Western frontier: both forms of property are "widely dispersed ... and therefore easier to steal" than fixed goods, and consequently both need to be protected by especially strong sanctions: "death to rustlers and horse thieves, trophy-ization for copyright infringers" (1994, 265). This is what Barlow fails to understand—despite the fact that he is a cattle rancher from Wyoming. For Barlow, technology determines how we define private property. For Jeter, to the contrary, it is struggles over property that determine which technologies we develop in the first place. Unlimited copying is now technically possible, but so is a system of tracking so precise, and so extensive, that not a single byte, in any machine, anywhere, will escape being identified and accounted for. Which one of these technical possibilities actually comes to pass is a matter to be determined in the marketplace, and in the courts. Economic rationality is the bottom line for Jeter, just as it is for Marxists and free-marketeers alike. Despite our supposed information glut, Jeter argues, the creation of valuable intellectual property is still a rare process. It is subject to the constraints of scarcity and Malthusian competition, the "immutable laws" of economics and physics. As any economist will tell you, there is no free lunch, for much the same reason that there is no perpetual motion machine. Indeed, Jeter says, "if stuff [is] worth buying and selling—not just hard physical stuff, but intellectual property as well—then it [is] worth stealing, too" (253).

House Rules. Appropriately, then, the biggest problem in Jeter's argument is a matter of economics as well. Intellectual property rights ought to work in the same way at all levels of the economy; Jeter says, "the same rule of survival applie[s] to big international corporations, to midlevel localized players and entrepreneurs, to scrabbling, scribbling little content creators" (251). But *Noir* depicts a world in which—just as in our own—such an equivalence cannot be made. In our heavily networked society, private individuals do not have the same rights as multinational corporations. Each individual may be a node in the network, but the corporation is the network itself. The more corporations are recognized as persons, as has increasingly been the case under United States law, the less unincorporated individuals are able to be so recognized. Under these conditions, corporate property and individual property are not the same thing at all. Corporations are not subject to "the same rule of survival" as individuals; their struggle is a Nietzschean one to increase their dominance, rather than a Malthusian/ Darwinian one just to survive. As Harrisch of DynaZauber muses at one point: "other people had to deal with win-or-lose situations; he'd made sure that his own contained no possibilities other than winning" (449). And Harrisch is right, even though he himself is killed at the end of the book. You may be able to beat Harrisch as an individual, but in the long run, you can't beat DynaZauber, just as you can't beat the house in Vegas. This irony is missing from Jeter's overt polemics about copyright, but it returns with a vengeance in the actual plot of *Noir*. McNihil, the novel's detective protagonist, captures someone whom he believes to be a copyright offender; but it turns out that, in doing this, he has really become the victim of an elaborate sting operation. He is then sent on a mission to recover DynaZauber's corporate intellectual property from theft, but it turns out that this, too, is really a false front for a sinister plot to extend corporate domination. At every turn, the battle for creators' rights ends up only benefiting Disney or DynaZauber.

The Mediasphere. Corporate domination is the issue here. We live in a world of images and sound bites. The electronic media are to us what "nature" was to earlier times. That is to say, the electronic media are the inescapable background against which we live our lives and from which we derive our references and meanings. Everyone now understands what Andy Warhol was perhaps the first to enunciate: our experience has to do not so much with fruits and flowers, or rivers and mountains, as with Campbell soup can labels and images of Marilyn and Elvis. These endlessly replicating icons are the very fabric of our lives. That is why appropriation, or sampling, is everywhere today: from rap songs, to films and videos, to prose fiction and installation art. Sampling is the best way, and perhaps the *only* way, for art to come to terms with a world of brand names, corporate logos, and simulacra. Pure originality is a myth, in any case; art and culture can only be made from previously existing art and culture. Even Jeter admits that, "bit by bit," every writer has "constructed the world inside his head from" previously existing texts by others (241). Indeed, Shakespeare would never have been able to write and produce his plays if copyright protection, as we define and enforce it today, had been in effect back in his time. Jessica Litman tells the story of how copyright law has evolved, in the course of the twentieth century, from something that balanced the rights of ownership with the rights of the public, to something that seeks to maximize the economic "incentive" to produce new works. The entertainment conglomerates claim that they need strong copyright protection if they are to find it worthwhile to continue to churn out their products. On the other hand, if they have any reason to fear that their video and audio recordings will be stolen, they say that they simply will not have the "incentive" to make them anymore. Without adequate copyright protection, content producers will have no reason to make art, and creativity will just wither and die. But of course, when it comes to actual artists, instead of corporations, the situation is precisely the opposite of this. Individuals and small groups rarely create art out of economic incentive. And nobody can create without materials to work with. When the very fabric of our culture is copyrighted and trademarked and placed in private hands, then creativity dries up altogether. Individual creators certainly want to get paid for their work. But these creators, more than anyone else, need free and easy access to previously existing media—the raw material of art—to make their work in the first place. The law of copyright, as it is currently evolving, means that only large corporations, or artists "for hire" in their employ, will be able to pay for the samples that they need. No individual will be able to afford the enormous royalty fees that are the prerequisite for getting hold of samples. Today, the samples that I am using for this book are still freely available to me, according to standards of "fair use," but in the not-too-distant future, they probably no longer will be, and the publication and dissemination of the text you are reading now will be illegal.

Deconstructing Beck. In early 1998, a group calling itself Illegal Art released its first CD: *Deconstructing Beck*.[19] The album is built entirely out of samples taken, without authorization or payment, from music by the alternative-rock icon Beck. The samples are manipulated electronically in various ways by a number of pseudonymous artists. Some of the pieces on the album work explicitly as witty commentaries on their source. Others change the music unrecognizably, breaking it into abstract formal patterns. But all thirteen tracks flaunt their own illicit status. Indeed, when it made the album available for purchase on its Web site, Illegal Art also made sure to notify Beck's record company, publicist, and attorney, in effect daring them to sue.[20] In terms of both its production and its distribution, *Deconstructing Beck* thus asks basic questions about ownership and copyright. Who owns the images and sounds that are all around us? What does it mean to own a sound or an image anyway? What are the implications of reproducing one? For that matter, how do we even delineate a single image or sound? Where does one end and the next begin? Given a pre-existing visual or sonic source, how radically must it be changed before it is turned into something new? Should the notion of authorship apply to images and sounds themselves? Or only to the uses to which those images and sounds are put? Or should it not be utilized at all? It's especially appropriate that these questions are being asked in relation to Beck, whose 1996 album *Odelay* married the sampling technologies of hip-hop to the disaffected sensibility of white alternative rock. For Beck's own music is largely made up of samples. He incorporates everything from hip-hop to country music into a quirky sound that suggests many genres but belongs to none. It has rough edges, but it

never seems confrontational or aggressive. Beck is pleas-
ingly idiosyncratic, but in a manner that freely acknowledges larger cultural trends. His wry, laid-back, off-kilter
observations exude a sense of slacker cool. His lyrics always seem to be on the verge of making sense, but they
never congeal into fully identifiable meanings. Beck seems
to relish being a sort of Bob Dylan manqué. He doesn't claim
to be a spokesperson for "Generation X," but his offhanded
refusal of such a role just makes him seem like a representative figure all the more. In all these ways, Beck exemplifies the culture of appropriation at its most apolitical
and benign. Of course, Beck can only do this because his
samples are paid for by Geffen Records, ultimately a subsidiary of Vivendi/Universal. And it's this privilege that
Deconstructing Beck seeks to expose, by translating his music
into more dissonant and abrasive forms. Illegal Art subjects
Beck to his own treatment, while pointedly dispensing with
his corporate safety net.[21]

The Cut-Up. As the contrast between Beck and *Deconstructing Beck* suggests, the practice of sampling can take
many different forms and has a wide range of implications
and meanings. At one extreme, appropriation can be seen
as a kind of complacent submission to the dominant form
of the commodity. One might thus characterize Beck in the
terms of Jameson's definition of postmodern *pastiche* as "a
neutral practice of . . . mimicry, without any of parody's ulterior motives, amputated of the satiric impulse" (17). At
the other extreme, appropriation can be seen to work as a
disruptive act of "culture jamming" or of what the Situationists called *détournement* (Lütticken). This is certainly
the intention of the people behind *Deconstructing Beck.* One
can find support for both sides of this argument in William
Burroughs's discussion of the cut-up technique, which he
invented with Brion Gysin. Cut-ups are a kind of predigital

form of sampling. You literally cut pages of written text in half and paste them together in new configurations. On the one hand, Burroughs says, cut-ups are valuable tools to scramble the dominant codes and to break down our pre-programmed associations. Cut-ups "establish new connections between images, and one's range of vision consequently expands" (Burroughs and Gysin 4). On the other hand, Burroughs later pessimistically worries that cut-ups have only limited efficacy since they still assume, and still serve, the viral replication of the dominant language: "the copies can only repeat themselves word for word. A virus is a copy. You can pretty it up, cut it up, scramble it—it will reassemble in the same form" (1981, 166). Burroughs, like a range of cultural critics from Adorno to Jameson and Attali, thus ends up suggesting that repetition in popular culture is a symptom of commodification and industrial mass production, testifying to our enslavement.

The Politics of Sampling. Whatever role postmodern sampling plays in postmodern culture generally, it has special, privileged position in hip-hop and black culture. Tricia Rose argues that rap uses sampling "as a point of reference, as a means by which the process of repetition and recontextualization can be highlighted and privileged" (73). That is to say, sampling in rap is a way of making new connections between the past and the present (and also, perhaps, between both of these and the future). Following James Snead, Rose notes that the "cut" or "break beat" isolated from an older song by the sampling process "systematically ruptures equilibrium"; yet at the same time, "the 'break beat' itself is ... repositioned as repetition, as equilibrium inside the rupture"(70). Of course, not all hip-hop samples are break beats, but even the ones that aren't tend to follow this double logic of rupture and renovation by means of repetition. Rose thus opposes the way that rap music affirms repetition through sampling to the Frankfurt School critique of commodified repetition (72). Appropriation in hip-hop is neither subversive nor conformist; it neither challenges the commodity form nor complacently celebrates it. Rather, hip-hop sampling is an affirmative practice, an exuberant act of reclamation and reconstruction. For instance, when Missy Elliott samples Ann Peebles's 1974 soul hit "I Can't Stand the Rain" in her 1997 rap song "The Rain (Supa Dupa Fly)," she revitalizes the older song by placing it in a new context. Peebles sings a lament for a man who has left her; Missy boasts instead of getting rid of a man "before he can dump me," as well as proclaiming her own "supa fly" hipness and her control over all aspects of the song's production. Ann Peebles's soulful complaint is neither negated nor ironized; rather, the way it is looped through the new song, as a refrain, allows it to take on new powers.[22]

Blond Ambition. All this can be contrasted to the way that Beck both samples and imitates black music on *Odelay* —and even more on his subsequent album *Midnite Vultures*, on the cover of which he presents himself in whiteface. By calling attention to his appropriations, by highlighting them and placing them postmodernistically "in quotation marks," Beck is able to signify doubly. On the one hand, he seems to be holding up each of his samples and saying: "listen to how much fun this funky guitar line is—enjoy it!" At the same time, he seems to be smirking as he points to each sample from a distance, as if to say: "you don't think I'd take such corny emotionalism seriously, do you?" In this way, Beck complicates, and raises to a higher power, the old story according to which white artists have so often ripped off black artists, getting rich by imitating the latter's musical innovations. In contrast to, say, Mick Jagger, Beck exculpates himself by freely admitting that he is stealing. Where

Jagger wants to convince his (white) audience of his own authenticity (which is to say, in this context, his pseudo-blackness), Beck instead ridicules black music and black musicians precisely for being, as it were, authentic—something that the über-white Beck himself is way too cool and self-conscious ever to be. This is the way that postmodern, ironic appropriation is different from plain, old-fashioned high modernist stealing. But one thing remains common to Jagger's and Beck's methods: both of them are projecting, and defining themselves against, an image of black music as being naively emotional and unself-conscious.

Beyond the Society of the Spectacle. It is only in a very particular and qualified sense that we still live in what Guy Debord long ago called "the society of the spectacle." For Debord, the spectacle is a global phenomenon of alienation and mystification: "All that once was directly lived has become mere representation. Images detached from every aspect of life merge into a common stream, and the former unity of life is lost forever" (12).[23] In this detachment, the spectacle is false consciousness and particularly "a false consciousness of time" (114). As for the message of the spectacle, it is so hilariously deadpan that it well could have been uttered by Andy Warhol himself: "All it says is: *Everything that appears is good; whatever is good will appear*" (15). Debord sees the spectacle in Manichaean terms, as the *ne plus ultra* of ideological mystification. But isn't it actually Debord himself who is deluded? There never was a time when life was "directly lived," instead of being diverted into representations; there never was a "unity of life," as opposed to the separation imposed by the detaching of images from their original contexts. The unity of a life directly lived is a fiction; it is something that you can only find in Hollywood movies and that never occurred to anyone before the movies were invented. Warhol knew this;

everyone who goes to the movies knows this. Debord is apparently the only moviegoer who still doesn't know that everybody else knows that movies are fictions. As for television, Warhol suggests that it is the great demystifier. Far from deluding us with false representations, television is precisely the medium that makes us most acutely aware that everything we see, hear, and feel is just a representation. It's not a question of finding the real behind all its distorted representations, but of realizing that these representations are themselves the only reality there is. And if this was already the case in Warhol's time, it is all the more so in today's post-Warhol world of cable and satellite television and the Internet. There is nothing like Debord's grand spectacle, no totalizing system of false representations that would masquerade as actual life. Instead, we have a plethora of tiny spectacles, each of which calls explicit attention to its own status of merely being a spectacle. Each spectacle is a monad, entirely self-contained and self-enclosed, yet connected over the network to all the rest.

In Memory of My Feelings. Andy Warhol tells the story of the death of his emotions: "During the 60s, I think, people forgot what emotions were supposed to be. And I don't think they've ever remembered. I think that once you see emotions from a certain angle you can never think of them as real again. That's what more or less has happened to me. I don't really know if I was ever capable of love, but after the 60s I never thought in terms of 'love' again" (1975, 27). As usual, what Warhol presents as his own experience has since become a common theme in general accounts of postmodern culture. Thus J. G. Ballard writes of "the death of affect" under the impact of high technology (1985, 1), and Fredric Jameson of "the waning of affect" as a result of the commodification of lived experience (10ff). But whereas Ballard and Jameson seem to deplore this situation, Warhol greets it with a positive satisfaction. He is happy to be free from the pretensions of emotion. He credits television for thus liberating him: "I kept the TV on all the time, especially while people were telling me their problems, and the television I found to be just diverting enough so the problems people told me didn't really affect me any more. It was like some kind of magic" (24). With the television on, no problem is desperate or urgent any longer. Whatever it is, it's just part of the flow. What's more, Warhol realizes that, like on TV, every emotional state is really just a performance. We do not spontaneously have feelings; rather, we are always trying to make our behavior conform to our pre-established ideas of "what emotions [are] supposed to be." We seek to play up our feelings, like in the movies, but everyday life in this highly technologized society actually has much more in common with the small screen: "The movies make emotions look so strong and real, whereas when things really do happen to you, it's like watching television—you don't feel anything" (91). And that's where Warhol gets his blank and disaffected persona, his self-described "affectless gaze" of "basically passive astonishment" (10). It's sort of like an updated, postmodern version of Kant's notion of aesthetic disinterest (Kant 1987, 45–53). For it is only when you don't care anymore about your immediate personal problems—as Warhol says television taught him not to care—that you are free to reflect instead on surfaces and appearances. Both in his calculated positioning of himself as a voyeur and in his choice of objects—knick-knacks, fashions, gossip, and behaviors—to be interested in, Warhol is an exemplary postmodern aesthete.[24]

Fascination. What comes after emotion, or after love? The most important thing about Warhol's story of the death of his emotions is that he doesn't regard this death as a tragic loss. He's glad to have been wised up about love, and resolves not to fall for anything so corny ever again. Warhol nonetheless adds a crucial qualification to his account: "However, I became what you might call fascinated by certain people. And the fascination I experienced was probably very close to a certain kind of love" (27). Warhol goes on to describe his odd relationship with a woman he calls "Taxi" (probably Edie Sedgwick). He finds Taxi unbearable: she is selfish and narcissistic, she takes too much speed, she suffers from bulimia, and she is a pathological liar. Worst of all, she refuses ever to take a bath. But Warhol can't keep his eyes off Taxi: "She had a poignantly vacant, vulnerable quality that made her a reflection of everybody's private fantasies. Taxi could be anything you wanted her to be... She was a wonderful, beautiful blank. The mystique to end all mystiques" (33). Warhol even watches Taxi while she sleeps: "She fell asleep and I just couldn't stop looking at her, because I was so fascinated-but-horrified. Her hands kept crawling, they couldn't sleep, they couldn't stay still. She scratched herself constantly, digging her nails in and leaving marks. In three hours she woke up and said immediately that she hadn't been asleep" (36). This is Warhol's voyeurism at its creepiest. It meets its match, however, in Taxi's exhibitionism. Even asleep, she still offers herself as a spectacle. Warhol's gaze is like that of Orpheus, or better, a vampire, impotently trying to cross the gulf between life and death, while Taxi is like a zombie, convulsively twitching with a semblance of life, even during the little death that is sleep. The exchange between this vampiric gaze and this zombified body is the play of fascination. Such is the postmodern, or posthuman, equivalent of love. It's a love in which the subject and the object never meet; even connected by a glance or a stare, they remain apart. Such love is a cool, aesthetic contemplation; it is even "indifferent to the existence of the object" being loved, which is Kant's definition of aesthetic disinterest (1987, 51).

Even Better Than the Real Thing. Edie Sedgwick was just the beginning. Today, the mystique of celebrity extends much further than it did in Warhol's time. Such is the subject of Bret Easton Ellis's novel *Glamorama*. The book is filled with epic catalogs of celebrity names from the mid-1990s: "Jonathon Schaech, Carolyn Murphy, Brandon Lee, Chandra North, Shalom Harlow, John Leguizamo, Kirsty Hume, Mark Vanderloo, JFK Jr., Brad Pitt, Gwyneth Paltrow, Patsy Kensit, Noel Gallagher, Alicia Silverstone, and someone who I'm fairly sure is Beck or looks like Beck" (179). Everything is surfaces and appearances. To be is to be perceived. Media coverage is the only proof of your existence. If you haven't been featured on MTV, you're nothing. Victor Ward, the narrator of *Glamorama,* wanders through an endless succession of chic restaurants and parties and clubs in New York, London, and Paris. His life is a continual blur of drinks and joints and Xanax, designer labels and photo ops and glances into the mirror. Victor cannot tell his life from a movie. Film crews seem to follow him everywhere, giving him scripts to perform and recording even his most intimate moments. Gradually, Victor falls in with a cell of terrorist supermodels. These beautiful people carry bombs in Gucci or Louis Vuitton bags and plant them in cafes, hotels, and airplanes. Their attacks are calculated media events, tautological rhetorical gestures, pure propaganda of the deed. They serve no deeper purpose than to demonstrate, over and over again, the supremacy of images and the tyrannical power of celebrity and beauty. We used to think that beautiful looks and surfaces existed on their own and that they were only later reflected in the mirror of the media. But now we know that there is no such reflection. For the surface of things and the surface of the media mirror are already one and the same. And beauty is the effect of this identity, not its cause. Following this fatal logic of appearances, Victor gets absorbed by his own image. As the novel proceeds, his identity disintegrates, becoming ever less stable and ever more tenuous. Finally, Victor splits in half. There is nothing left of him but a pair of echoing voices. On one side, he is replaced by his own idealized double; he turns into the star of his own movie, the fulfillment of everyone's fantasy of what he ought to be. On the other side, he is reduced to the empty, discarded husk of himself, left behind when the set was struck and the film crew moved elsewhere.[25]

Fifteen Minutes. Warhol's best-known statement is doubtless the one about how, in the future, everyone will be famous for fifteen minutes. This is usually taken as a comment on the way that our culture is so obsessed with fame and with celebrities. Today, nobody is surprised to hear that "celebrity lifestyles are a collective fantasy," or that celebrity "has to be manufactured," and that it "requires serious effort to be sustained" (Sardar).[26] Now, Warhol did as much as anyone to make ideas like these into commonplaces. The pursuit of fame is central to his life and art alike. A celebrity himself, he was also an ardent observer, and a shrewd promoter, of celebrities. But there is more to Warhol's statement than this. For one thing, there is the suggestion that, not just the talented or lucky few, but everyone will be famous someday. Warhol's attitude is genuinely democratic; he is not an elitist about celebrity. He simply takes it for granted that, in a society dominated by the media of mechanical and electronic reproduction, celebrity is a fundamental mode of appearance. This can even be stated as a syllogism: to exist is to be visible, to be visible is to be famous, and to be famous is to be beautiful. Thus Warhol insists, "I've never met a person I couldn't call a beauty. Every person has beauty at some point in their lifetime. If everybody's not a beauty, then nobody is" (1975, 61–62). At least, that is one side of the story. The other side

is the part about fifteen minutes. Warhol knows that beauty
never lasts. Fame, too, is fleeting and uncertain. Even the
greatest celebrities are bound to be forgotten once their
fifteen minutes are up. The media do not linger over the
past; they are always moving restlessly onward, in search
of the Next Big Thing. Fame is just a fiction; the image of a
celebrity is a manufactured ideal that no flesh and blood
person can ever actually live up to. Warhol explores this
irony in his endlessly repeated portraits of Marilyn Monroe
and other media icons.[27] And he makes a similar point in
his *Diaries*, with their flat, mundane descriptions—blem-
ishes included—of all the celebrities he encountered.

Almost Famous. Spider Jerusalem is the hero of *Transmetropolitan*. Spider is bald, lanky, and heavily tattooed,
with an acerbic prose style, a love for high-tech weaponry,
and a devotion to all sorts of pharmacological excess. He is a
muckraking political journalist, sort of like Woodward and
Bernstein combined with Hunter S. Thompson. Spider has
a perpetual love/hate relationship with the City. Its sensory
overload, technological overkill, and sheer human density
drive him crazy. But the City's extremes are also the only
stimulants powerful enough to fuel his writing. Spider is a
postmodern *flâneur*, restlessly wandering the City's streets
in search of a story. He generally finds more than he has
bargained for: everything from police-sanctioned killings
and cover-ups to political scandals involving cloning, child
abuse, and real estate deals. When it all becomes too much,
Spider holes up in his bunker and tries to stupefy himself
with drink, drugs, and TV. But this never works, because
Spider cannot escape his own image. The problem is his
newspaper columns have made him famous. Nearly everyone in the City knows who he is; they love his identity as a
crusading journalist with a gonzo lifestyle, even as they ignore his political messages. It's hard to "speak truth to

power" when your very act of speaking is being marketed as entertainment. And so Spider suffers existentially, even though he profits financially, from being a celebrity. In one issue of *Transmetropolitan* (Ellis and Robertson 2002, 5–26), Spider's assistants find him cowering in a corner, waving around a gun and muttering paranoid threats. What drives him over the edge is not just too many drugs, but also seeing himself portrayed on TV. Spider appears on the tube as a cuddly cartoon character for kids, as a macho action figure, and as a stud in a porno flick. In the comic, each of these visions is drawn in a different style by a different cartoonist. The visual diversity highlights the schizophrenic meltdown of Spider's personality. Eventually, Spider can only imagine himself in the form of one or another media stereotype; in a drugged-out stupor, he hallucinates, first his vengeance as an all-powerful superhero, then his martyrdom as a lynching victim torn apart by his own fans. There's no easy solution to this dilemma. Spider needs his fame in order to get his message out, but this fame guarantees that the message will be distorted. At the end of the issue, Spider tells his fans that he didn't want them "paying attention" to him personally; all he wanted was for them "to *hear* me!" To which the fans reply: "We did. We just didn't listen" (Ellis and Robertson 2002, 26). As Spider already knows but is forced to learn again, no message can control the conditions of its own reception.

Exposure. The first Web cam, the Trojan Room Coffee Machine, from the Computer Laboratory at the University of Cambridge, went online in 1991.[28] It showed a tiny image of a coffee pot. The image was refreshed about three times a minute. The cam served the pragmatic purpose of letting researchers elsewhere in the building know when fresh coffee was available. Only later did it become famous as the first real-time, continually updated image to be made available on the World Wide Web (Stafford-Fraser). It took another half decade before anyone had the idea, or the courage, to turn a Web cam upon him- or herself. Jennifer Ringley's JenniCam[29] went online in 1996; ever since, it has broadcast images, refreshed every minute, of her day-to-day existence. Over the years, Jenni has occasionally played to the camera, and the cam's notoriety has made her into a minor celebrity, which has affected her in various ways. But for the most part, Jenni has just gone on with her life, much as she might have done without the cam. Originally, she says in the FAQs on her Web site, the JenniCam was merely "intended to be a fun way my mom or friends could keep tabs on me, and an interesting use for the digital camera I bought on a whim in the bookstore. I never really contemplated the ramifications of it." Now, years later, the cam has become a habit: "I keep JenniCam alive not because I want to be watched, but because I simply don't mind being watched." Indeed, there really isn't all that much to see. The images are generally bland and humdrum; the same goes for the entries in Jenni's written diary. Jenni's life is not much more interesting to outsiders than the question of whether the Trojan Room Coffee Machine is full or empty. Even Jenni's moments of high melodrama—like when she and her best friend's fiancé fell in love, leading to said fiancé's dumping said friend and moving in with Jenni instead—end up seeming fairly routine.[30] I don't mean to imply that Jenni is a boring person; she's no worse in this regard than anybody else. It's just that what the JenniCam shows is unimportant, compared to the sheer fact that it is always on, so that Jenni's life is presented to the world in its sheer ordinariness. To be is to be perceived.

Don't Look Now. We've come a long way since the Trojan Room Coffee Machine, and version 1.0 of the JenniCam. Today, Web cams are as plentiful as surveillance cameras:

there are far more of them than anyone could possibly watch. There are cams for every conceivable taste, purpose, and interest. Sheriff Joe Arpaio's Live Jail Cam, direct from the Madison Street Jail in Maricopa County, Phoenix, Arizona, is supposed to work as a deterrent to crime.[31] "We Live in Public," showing the daily life and sexual exploits of an Internet entrepreneur, was an expensive publicity stunt.[32] The Kathicam[33] focuses on a naive young woman who continually bemoans the fact that guys are surfing over to ESPN's Web site instead of watching her; of course, the site is actually run by ESPN itself, and the main character is played by an actress (Elliott 2001). There are even Web cams belonging to precocious fourteen- and fifteen-year-old (or so they claim) girls, who use them to flaunt their pubescent bodies,[34] not to mention secondary Web sites that compile stills from such teenage girls' cams, presumably for the delectation of pedophilic voyeurs in search of images to drool over.[35] Though a few Web cams actually do work as surveillance devices—in one case, a cam even caught burglars in the act (Higgins)—most of them do not serve any such purpose. If anything, they point to a dearth of surveillance. There are too many real-time images and not enough real-time eyeballs to watch them. Of course, some popular and profitable Web cams do exist, like the porn sites that charge on a pay-per-view basis.[36] But the vast majority of cams cannot sell their images or even get people to look at them for free. It is not a question, therefore, of Panoptical surveillance, but rather of an over-the-top performative exhibitionism. Even the most modest and unstaged Web cam is—by its very existence—clamoring for attention. Every Web cam incessantly repeats the same monotonous exclamation: "Look at me! Look at me!" Here, the medium really is the message. As McLuhan says, "the 'message' of any medium or technology is the change of scale or pace or pattern that it introduces into human affairs" (1994, 8). In the network society, people make contact over vast distances; I am connected to everyone, but in proximity to no one. In this situation, if I do not take steps to make myself visible, the chances are that I will just disappear. My very existence is staked upon my Web cam. I must continue my performance, even if nobody is watching it.

Family Values. The Web site of Stuart Tiros III is vaguely creepy and disquieting, in a way that many personal Web sites are.[37] It consists of family photos, written reminiscences, and lists of favorite movies and songs. In pictures and words, Stuart tells the story of his life. We see him at birthday parties, on family vacations, playing the guitar, and heading off for college. Stuart grew up in a typical American white suburban middle-class household. His father was an engineer for NASA; his mother stayed home with Stuart and his sister. Today, the sister is married, and has a child. Stuart himself has been married and divorced; he makes his living installing surveillance cameras in convenience stores. All in all, Stuart's Web site is as painfully earnest in its feelings as it is banal and unrevealing in its content. Many of its details are predictably generic: after all, what white suburban kid who grew up in the 1980s wouldn't list *The Breakfast Club* as one of his favorite movies? But other, quirkier details are left unexplained. For instance, why is Stuart so attached to that garish green wallpaper? And why is Dad in drag in one picture? Indeed, most of Stuart's photos are ever so slightly off. They are too ordinary to be classified as idiosyncratic. Yet they have an edge of kitschy hyperreality, a too-muchness that isn't quite right. In one photo, for instance, the young Stuart is wearing a T-shirt that reads "I Love DAD." He looks toward the camera, smiling just a bit too earnestly. It is like one of those fake smiles we see on TV. His little sister, next to him, makes an exaggerated grimace. Presumably she is doing this as a joke, but her expression seems overly awkward and strained. Images like this are embarrassing, not because they reveal something unsavory, but precisely because they do not reveal anything at all. This is American suburban family life in all its vapidity. But the banality of everyday existence resists being made public. You cannot observe someone's intimate experience without violating it. These photos are creepy, because they fail to bridge the gap between private and public. Just looking at them feels like an intrusion, even though Stuart has freely and eagerly offered them to our gaze. Stuart wants to share his memories with the world, but he fails to understand that these memories have no meaning or resonance for anyone other than himself. Stuart Tiros's photos are disturbing, then, less because of what they tell us than simply because of the sheer fact that they are online. For the Web site is Stuart's sole proof and verification of his own existence: I have a home page, therefore I am.

Prosthetic. But there is more to this story. Stuart Tiros, it turns out, is a fictional character, and his Web site is a fabrication. It is actually part of a multimedia installation *Prosthetic* by William Scarbrough.[38] Besides Stuart's photos, presented in various formats, and live video surveillance feeds, *Prosthetic* also includes a series of audiotaped interviews. In these tapes, Stuart tells a very different story from the one recounted on his Web site. It turns out that Stuart Tiros III is a lonely, embittered man. As the result of an auto accident, he is confined to a wheelchair and must wear a prosthetic replacement for his damaged nose. The Web site photos are also a kind of prosthesis: a proxy for his shattered identity. Stuart claims to have found them in his basement; he doesn't know whose pictures they really are, but he finds them beautiful. In their quotidian banality, they suggest an ideal of normalcy that he has never possessed. And so Stuart claims the photos as his own. On the Web, he projects his own story into them. He invents a new, improved version of his life, one in which things happen the way he wishes they had. Behind the shelter of this fictive online identity, he meets people and even falls in love. But everything falls apart when Stuart's cyberlover tries to meet him in the flesh. Not wanting her to know of his disfigurement, he sends her away without a sight or a word. Stuart is pathetically unable to leap from the virtual to the actual, from the safety of his prostheses to the vulnerability of the here-and-now. But isn't this the general condition of virtual existence, of what Castells calls "the culture of real virtuality" (2000b, 355ff)? William Scarbrough creates Stuart Tiros as his fictional surrogate, who in turn creates another fictional surrogate to live his life for him. Yet in this context, fictional need not mean unreal. After all, to be is to be perceived—and vice versa. Stuart's Web site is as visible and accessible as any other; you can send him e-mail, and you may even receive an answer. It is just that the virtual realm seems to operate according to a logic of infinite regress. There is no pregiven, natural reference point, no first or final term. Everything virtual is a hyperlink; everything is a reference to, or a prosthesis for, something else. Images and narratives proliferate endlessly over the network, in a vain attempt to populate the void.[39]

Esse Is *Percipi*. George Berkeley is probably the philosopher in the Western tradition who most fully anticipates our current ideas about virtual reality. He notoriously argues that *esse* is *percipi*: to be is to be perceived. In his own time, Berkeley was merely taking the doctrine of empiricism to its logical extreme. If our minds contain nothing but atomistic perceptions—which is to say, *ideas* or representations—then it is superfluous to posit, in addition, a material world out there that would be independent of these ideas, although supposedly giving rise to them. Mental representations themselves are enough, says Berkeley, especially since—according to our initial assumptions—we can never get beyond them in any case. Berkeley's argument reads like an unintended *reductio ad absurdum* of what Richard Rorty calls "the 'idea' idea": the Cartesian notion that the mind is like a theater in which consciousness is a detached spectator that contemplates and manipulates special objects of inner sense (ideas or representations). The major philosophical question then becomes that of how our mental representations relate to their corresponding objects in the material world. The genius of Berkeley is to simply short-circuit this whole dilemma, by negating the material world altogether. His radical conclusions follow logically and powerfully from his dubious initial premises. Now, nearly all of the important twentieth-century philosophers reject "the 'idea' idea" in the first place.[40] But the representationalist approach remains alive and well in other fields, most notably in AI (artificial intelligence) research and in cognitive science. Cognitive scientists start from the assumption, not that computers should be understood by comparison to human minds, but rather that human minds themselves can already be understood in terms of computers. This is more than just "some rough analogy," says Andy Clark; "it is not that the brain is somehow *like* a computer," but that it "actually *is* some such device" (7–8). This means that cognitive scientists conceive minds, on the model of digital computers, as information processors that work by performing logical operations upon internal representations of external phenomena.[41] This is why Berkeley is still relevant today. In twenty-first-century terms, his argument may be rephrased as the claim that our experience is always already virtual. And that is indeed what the cognitive scientists say. They claim that the "real world" of our perceptions is in fact largely a construction of our own inner cognitive processes. "You and I, we humans, we mammals, we animals, inhabit a virtual world . . . the brain works as a sophisticated virtual reality computer" (Dawkins 1998, 275, 278). Our sense of reality is the product of a simulation. It only remains for the cognitive scientists to follow Berkeley all the way and jettison the "outer world" altogether as an extravagant, unnecessary hypothesis. There will then be no escaping the control of the network.

Eyes Wide Shut. Berkeley is surprisingly unperturbed by the obvious objection that, if his theory were correct, then objects would cease to exist whenever we stopped looking at them. He dismisses this worry on a number of grounds, all of which apply just as well to our current conceptions of virtual reality. First of all, Berkeley says, if things exist only as perceptions or representations in the mind, then that tree yonder just as surely exists when I am thinking of it with my eyes closed as it does when I am looking at it directly. In either case, the tree is being perceived as an idea by my inner sense. Berkeley brackets the whole question of the cause of perception; what matters is only its effect within my mind. But this is the very principle of virtual reality; as Deleuze puts it, "simulation designates the power of producing an *effect*" (1990, 263), even in the absence of anything that is supposed to be a cause. In the second place, Berkeley says, the fact that I may only perceive a given object intermittently doesn't impugn the consistency of the object. There is no reason why the object shouldn't have the same features and appear in the same place whenever I do happen to perceive it or think of it. There is no more reason for me to worry that the tree will be uprooted because I look away from it than there is for me to worry that the objects of a virtual world, or the icons on my desktop, will dissipate because I turn off my computer. In either case—when I look back at the tree, or when I turn the computer on again—I will find that everything is exactly the way I left it. In the third place, and most importantly for Berkeley, just because a given idea is no longer present in my mind does not mean that it is likewise absent from all other minds. The proper logical conclusion from the intermittency of my ideas is "not that [sensible objects] have no real existence," but rather that "*there must be some other mind wherein they exist*" (201). Ultimately, for Berkeley, this other mind is God. Today, we are more likely to say that it is the computer, or better, the network, on which the virtual reality simulation is being run. Perhaps this is the reason for the proliferation of home pages and Web cams. To be online is already to be perceived. Even if no one ever visits your Web site, you are still visible to the network itself. For the network is God, the unsleeping omni-voyeur. The *cogito* of virtual reality therefore reads: I am connected, therefore I exist. Long before the Internet, Warhol already understood this logic. His film *Empire* (1964) shows the Empire State Building in a single continuous stationary shot that lasts for over eight hours. Warhol's stated purpose in making this film was to turn the building into "a star!" (Smith 153). And we must say that he succeeded, just by virtue of having created the film. Nobody actually has to watch *Empire* in order for the movie to have its effect. As long as the film is rolling through the projector, the virtual, simulacral image is perceived, as it were, by the cinematic apparatus itself; and so the Empire State Building actually is a star.

Saving the Appearances. Perhaps the oddest thing about Berkeley's argument is his claim that, in fact, the argument has no pragmatic consequences. "After having wandered through the wild mazes of philosophy," he writes, we "return to the simple dictates of nature," and "come to think like other men" (158). Berkeley indulges in metaphysical speculation, the better to put an end to such speculation. He denies the existence of matter, he says, only in order to refute skepticism and vindicate the assumptions of common sense. This may seem like a crazy, and outrageously backward, way to proceed, but Berkeley's point is that the best way of "saving the appearances" is to show that there is nothing besides appearances, no real world behind this apparent one. In an immaterial world—or what is the same, a virtual world—nothing is hidden, and everything is precisely what it seems. In a certain way, then, Berkeley anticipates Nietzsche's polemic against those metaphysicians who distrust the senses. Berkeley could easily say, along with Nietzsche, that the senses "do not lie at all...The 'apparent' world is the only one; the 'real' world has only been lyingly added..." (1968a, 36). Berkeley's critique of skepticism is also oddly congruent with Nietzsche's critique of nihilism. For Berkeley, skepticism arises when we posit the existence of an external, material world, only to discover that we can know nothing about such a world and that we can have no access to it. For Nietzsche, similarly, nihilism arises when we posit the existence of a transcendent "real world," only to discover that such a world is empty and that we can have no access to it. Of course, it is crucial that Berkeley denies the existence of transcendent materiality, while Nietzsche denies the existence of transcendent ideality. The radical conclusions Nietzsche draws from his arguments could not be further from the pious conclusions Berkeley draws from his. For Nietzsche, everything changes when we learn to accept appearances; traditional values crumble, and everything must be created anew. For Berkeley, in contrast, nothing changes; the order of the world is confirmed, once we realize that everything is just an appearance. We can read Nietzsche and Berkeley, therefore, as rival science fiction writers, offering alternative visions of what Michael Heim calls "the metaphysics of virtual reality."

What Is Mind? No Matter. Early in the development of the Internet, John Perry Barlow celebrated cyberspace as "the native home of Mind" (1994). Barlow went on, just a few years later, to proclaim the independence of cyberspace, not only from the United States government and other nation-states, but from any material reality whatsoever: "cyberspace consists of transactions, relationships, and thought itself, but it is not where bodies live" (1996). What's crucial here is the separation of transactions and relationships from the bodies that give rise to them and that enter into them. We may well live in chains, but once we go online, our thoughts are free. Addressing the powers of the State, Barlow even divides physical from virtual authority: "We must declare our virtual selves immune to your sovereignty, even as we continue to consent to your rule over our bodies." Needless to say, Barlow is not very clear on how this division can be maintained. After all, if the cops physically drag me away from my keyboard, where does that leave my virtual self? For that matter, what if an online transaction goes bad, and the phone company responds by physically shutting down my network link? But Barlow floats loftily above all such contingencies. He combines an idealism worthy of Berkeley, with an all-American optimism straight out of Emerson. Not only is cyberspace the native home of Mind, but it even grows by "an act of nature," so that Barlow warns governments not to think that they "can build it, as though it were a public construction project." In fact, the Internet *was* originally built as a public construction project, financed by the United States Department of Defense. But why squabble over technicalities? Today, Barlow's vision and rhetoric may seem naive, but this is really a testimony to the success of his vision. We live in a world in which simulation is ever more taken for granted, and life online has become a banality. The Internet and the World Wide Web are no longer places for pioneers to explore and stake their claims; they have been absorbed into the texture of our everyday life. If Barlow's exceptionalism with regard to cyberspace seems dated, this is simply because virtual reality is no longer an exception: today, it is everywhere and everything.

What Is Matter? Never Mind. Today, in spite of Berkeley's and Nietzsche's admonitions, we do not trust appearances. We get uneasy, and suspicious, when we learn that so many things are simulations. Disney's California Adventure™ is a simulation of the State of California, but isn't California itself already a simulation of the American Dream, which in turn is itself a simulation? We want to know what is behind all the virtual reality displays that we seem to be living in. Who made them and for what purposes? Can it be that our bodies are actually sheathed in womblike, gelatinous cocoons, while our minds are stimulated with fake sensations, so that the network can harvest our psychic energy, as depicted in the film *The Matrix?* Such is the world of the Evil Genius of Descartes: "I will suppose therefore that not God, who is supremely good and the source of truth, but rather some malicious demon of the utmost power and cunning has employed all his energies in order to deceive me. I shall think that the sky, the air, the earth, colours, shapes, sounds and all external things are merely the delusion of dreams which he has devised to ensnare my judgment. I shall consider myself as not having hands or eyes, or flesh, or blood, or senses, but as falsely believing that I have all these things" (15). That is to say, everyday life is already virtual. Descartes only proposes this fantasy as a thought experiment; he advances it in order to disprove it, thereby establishing beyond all doubt the existence of the material world and the veracity of the Word of God. But today the Evil Genius is alive and well; the hypothesis could easily be a motto for any number of novels by Philip K. Dick, as well as for *The Matrix*. Berkeley argues that experience is virtual or mental, and therefore real. What could be more real, after all, than the ideas in my own mind? As Descartes already said, even the Evil Genius cannot stop me from thinking them. But Neo (Keanu Reeves) in *The Matrix* is taught precisely the opposite. The crucial scene is the one where Neo learns how to bend a spoon with the power of his mind. According to Berkeley, such a spoon must be real, because it is a mental representation. But for Neo, since the spoon is an idea in his mind, it follows that "there is no spoon." And that is why he can do whatever he wants with it. In short, Neo's omnipotence is built on a vacuum. Žižek is right to object that in this scene "the film does not go far enough." If Neo rejects the reality of the spoon, then he ought to go all the way with his rejection: as Žižek says, his "further step should have been to accept the Buddhist proposition that I MYSELF, the subject, do not exist" any more than the spoon does (1999).

The Ghost in the Machine. How does the network control us? In *The Matrix*, it dazzles us with a virtual reality simulation that is so complete, and so compelling, that we cannot help believing in it. As Morpheus (Laurence Fishburne) tells Neo, the truth is "that you are a slave ... That you, like everyone else, was born into bondage ... kept inside a prison that you cannot smell, taste, or touch. A prison for your mind."[42] The problem is less that we are in prison than that we do not realize that we are. For this is a prison without bars and walls; indeed, it is entirely undetectable. The simulation we live in is infinite and unbounded;

in Kantian terms, it is coextensive with the field of all possible experience. What might count as liberation in such a context? By definition, nothing that we do *within* the simulation is of any consequence. Freedom can only mean that we reject virtual reality altogether, refuse the consolations of illusion, and awaken to what Morpheus calls "the desert of the real." It's a harsh, powerfully gnostic vision: all that remains of the "real" world is a devastated, nearly uninhabitable wasteland. There's nothing there, no redemption or solace. Once you've attained the real, there's nothing left to do except go back into the Matrix and carry on the impossible fight against it. But this is where the film tries to have it both ways. It wants to condemn simulation and at the same time profit from it. For what Neo learns in the course of his training is not so much how to escape from virtual reality as how to use it to his own advantage. What makes the film so cool, and so compelling, is stuff like Neo's ability to bend spoons with his mind, to make those incredible leaps and martial-arts kicks, and even to come back to life after he's been shot. Neo transforms the world into a giant computer game: virtual reality raised to a higher power. *The Matrix* thus celebrates false appearances, under the guise of condemning them. It reaffirms the very logic of simulation that it identifies as the source of our oppression.[43] We might do well to say, updating Nietzsche: "I fear we are not getting rid of the Matrix because we still believe in special effects."[44]

The Big Picture. "Audrey felt the floor shift under his feet and he was standing at the epicenter of a vast web. In that moment, he knew its purpose, knew the reason for suffering, fear, sex, and death. It was all intended to keep human slaves imprisoned in physical bodies while a monstrous matador waved his cloth in the sky, sword ready for the kill" (Burroughs 1981, 309). In this passage from *Cities of the Red Night,* Burroughs offers a radically different account of our enslavement from the one we get in *The Matrix.* For Burroughs, virtuality is not the problem. The Real itself, in all its materiality, is what imprisons us. Malevolent intention, or purpose, knits all the disparate strands of reality together into a vast network, within whose toils we are trapped like flies in a spider's web. We have been selected as sacrificial victims of this malevolent power so that our destruction may bear witness to its glory, just as slain bulls attest to the virile magnificence of the matador. In these conditions, it is not the prison of the mind that we have to fear, but far worse—the prison of the body. (I mean this phrase in both senses of the genitive: the prison that the body itself is, and the prison in which the body is confined.) The power that enslaves us is not concerned with what we think, but with what we do. More precisely, the very effect of this power is to separate mind from body, thought from action, and the virtual from the real. It is to turn us into Cartesian subjects, impotent ghosts trapped in clunky "survival machines." Thinking is evacuated from the body and banished to the realm of the merely hypothetical. And that is what it means to be "imprisoned in physical bodies." The salvation that *The Matrix* promises us is not any sort of freedom, for it is still premised upon this separation, this imprisonment.

Body and Soul. Theologians have long maintained that the body is a prison, and the soul is its prisoner, freed only at the moment of death. But Foucault suggests that the actual condition of modern humanity is rather the reverse: "the soul is the prison of the body" (1979, 30). That is to say, the body is the prisoner, and the soul is both its jailer and its jail. This imprisonment is not oriented toward the final dissolution of ties in death; rather, it is renewed at every moment, throughout life. The soul determines—and enforces—the limits of what the imprisoned body can do. And this is not just a metaphor. Foucault insists that the soul is perfectly real, even if it is "non-corporal" (29). Think of it as an induced informational pattern, or better, as a field of forces and intensities. My soul is the virtual portion of my being: the set of all the constraints, and conditions of possibility, that come together to make me who I am. This is something that runs deeper than "personality" or "character." For as Foucault puts it, "the man described for us, whom we are invited to free, is already in himself the effect of a subjection much more profound than himself. A 'soul' inhabits him and brings him to existence, which is itself a factor in the mastery that power exercises over the body" (30). Even when I have freed myself from all external control and realized my full inner potential—as Neo does in *The Matrix*—my freedom is still premised upon a deeper subjection. For the soul whose powers I unfurl, and whose dictates I fulfill, is not really anything of mine. It just "inhabits" me, and animates me, like a viral infection or a demonic possession. Against the traditional dualism of soul and body, Foucault proposes a subtler and more complex relation. The soul "is not a substance" any different from the body, he says, but an "element" through which social and political "technologies of power" affect the body and within which the body itself moves. The soul is the actual, physical body's virtual aura—its noncorporal double—that envelops it from without, and impels it from within, without ever being reducible to identity with it.

Ontology of the Virtual. In novel after novel, Philip K. Dick explores the paradoxes of shifting levels of reality. All that is solid melts into air. What seems to be real in one chapter turns out to be an illusion or a fictional construction in the next. Familiar places and environments dissolve to reveal sinister underpinnings. Identities shift, and characters metamorphose into strange and alarming shapes. Each level of appearances is discredited in turn by what seems like a tear in the very fabric of reality. Sometimes these shifts can be explained away as schizoid breaks on the part of one or another character. Dick's protagonists themselves often wonder if they are going crazy. But in the long run, these disruptions go beyond individual psychology. At best, they can be explained as the productions of something like a Cartesian Evil Genius.[45] But even this may be giving them too empirical a basis. Unlike *The Matrix*, Dick's fictions never encounter "the desert of the real." They never get beyond appearances, never reach an ultimate level of reality. In Dick's late theological novels,[46] there are signs and messages that *may* have come from a transcendent Beyond. But in most of the earlier books, we don't even get that. And Dick is well aware, in any case, that signs and messages are always ambiguous, always subject to the vagaries of interpretation. This is why most readings of Dick have focused on issues of epistemology and hermeneutics. Dick is commonly said to be asking questions about what is real, what is authentically human, and how we can ever know for certain about such things. The most acute readings, such as those of Katherine Hayles (160–91), Pamela Jackson, and Carl Freedman (1984), ask metaquestions about these more basic questions. Freedman, for instance, argues that the obsessive interpretive dilemmas of Dick's novels reflect the ways that a paranoid hermeneutics is inscribed in the logic of capitalism itself: a delirium of interpretation is generated both by the mysteries of commodity fetishism and by the actually conspiratorial nature of corporate intervention in the political process. Valuable as such readings are, however, I think that they miss a crucial aspect of Dick's investigations. I would like to suggest that Dick's concerns are ontological before they are epistemological or hermeneutical. That is to say, Dick's novels are primarily about the *being* of simulation and virtual reality. They explore the mode of existence of various states of semireality and unreality: like being dead, or being a machine, or being cut off from God, or being a drug addict, or being lost inside a fantasy world of suburban stereotypes, or being overwhelmed by media images. What is it like to live in a simulation? How does it feel to move among multiple, mutually contradictory appearances? What are the actual qualities of virtual existence? For Dick, Being is not a plenitude. It is always somehow fake, or trashy, or incomplete, or unstable, or radically inconsistent. And Dick's novels describe, in excruciating detail, the lived experience of this unreality, or not-quite-reality, that is yet not simply absence or nonexistence.[47]

Money Is a Kind of Poetry. Philip K. Dick's novel *Ubik* places money at the heart of virtual existence. In the future world of the novel, household objects are endowed with sentience and apparently have legal status as subjects. They demand payment each time they are used: a nickel to open the door (23–24), a quarter to use the shower (31), a dime to get food out of the refrigerator (32). When the protagonist Joe Chip tries to leave his "conapt" without paying, the door threatens him with a lawsuit (24). Even though Joe is always broke, he is still convinced of the necessity of this system. He defends it in terms that a libertarian free-market economist like Steven E. Landsburg would most likely agree with: being able to take a shower for free, Joe says, is "not economically feasible. How can it operate on that basis?" (31). Money is evidently the glue that holds the world together; without it, things would stop working, and appearances would decay. Indeed, this is what happens in the second half of the novel. Everything breaks down after the apparent murder of Joe's boss, Glen Runciter. The money goes first: coins that used to be legal tender suddenly revert to "obsolete" forms of "merely numismatical interest" (88). This trouble with money is only the leading edge of a more general process; soon everything is "slowing down into entropy" (119). Food and cigarettes become instantaneously stale; people lose vitality, aging and dying suddenly. Objects devolve into earlier versions of themselves: the TV set regresses into an "oldtime AM radio" (132), and so on. The only good that comes of these changes, as far as Joe can see, is that the old-fashioned coffeepot he finds in his kitchen "lacked the coin slot, operating obviously toll-free" (131). Fortunately, the decay process does not go unchecked. A mysterious counterforce arises to oppose it. But this revivifying force is also channeled through money. Runciter's image starts appearing on coins, and "a beautiful steel-engraving portrait" of him adorns the bills (106). At least this "Runciter money" is of recent issue, though there is some debate as to whether it is "real," in the specific sense of being acceptable legal tender (106). The mystery is compounded at the very end of the novel, when in a final twist Runciter himself finds "Joe Chip money" in his own pocket (216). The book passes through a dizzying succession of possible worlds, all of which are more or less virtual. In every case, money is the index of the world's degree of reality. This is why the novel focuses so obsessively on coins and bills, as opposed to checks, credit cards, or other, more abstract, ways of making payments and signifying wealth. Joe Chip is hopeless at managing money; he always pays (or borrows) in cash because his credit line has been shut off (23). But the phenomenon goes beyond Joe. Of course, coins and paper money are themselves virtual: their value is conventional and symbolic, rather than intrinsic. But as tokens, coins and bills are more concrete, more fully physical, than any other financial instruments. You can feel them between your fingers and look at the pictures on them. In *Ubik*, this materiality of the currency is what grounds and makes possible the insubstantiality of everything else. The engraved heads of George Washington, Walt Disney (who is "supposed to be on the fifty-cent piece" [105] in the novel's future world of 1992), and Glen Runciter preside over the play of simulacra.

Half-Life. After his apparent death, Glen Runciter manifests himself in Joe Chip's world in a number of different ways, besides appearing on coins and bills. Joe hears Runciter's voice on the telephone (94). He finds messages from Runciter on matchbook covers (103), on parking tickets (161), and in bathroom graffiti (120, 123). He even sees Runciter on TV, addressing him directly (127). Puzzling over these apparitions, together with his experiences of decay, Joe concludes that he himself is in fact dead, and Runciter is the one who is still alive. The dead in *Ubik* linger on in "half-life," a condition of intermittent cerebral activity, while their bodies are cryogenically frozen (11). In this state, the living can still talk to them—at least for a time. Half-life gradually ebbs away, and finally diminishes to zero; only then are the dead definitively gone. (It is no accident that Dick uses a term that originally referred to radioactive decay.) Joe Chip experiences this half-life from the inside. He is relegated to a world that is self-contained but limited and mortal. Things revert to earlier forms and crumble into dust. The half-life world's appearances are all-embracing so that it is impossible to get beyond them, but somehow they *feel* incomplete. This is as good a definition of virtual reality as any. Something is indeed missing from Joe Chip's world, but this *something* is not anything in particular; certainly, it is nothing that he can name. For what is missing is not a *thing* at all. What is missing, rather, is a way out, an opening onto otherness. The half-life world is rigorously self-enclosed: a place of mere appearances, *phenomena* in a Kantian sense. There is no access to any Outside. And yet, the inhabitants of this half-life world are obliged to believe that such an Outside (like Kant's realm of *noumena*) must somehow exist, even though they cannot actually reach it. As the novel proceeds, Joe Chip realizes that he is caught in the middle of a Manichaean struggle between the forces of entropy and of redemption. Entropy is embodied by Jory, a creepy adolescent boy who cannibalistically devours the other dead. Redemption is bizarrely embodied by Ubik, an aerosol spray can "with bright stripes, balloons and letter-ing glorifying its shiny surfaces," (181) that works as a "re-ality support" (127). Although these opposed forces are cosmic in scope, they are both immanent to the half-life world. Indeed, everything that Joe Chip encounters "origi-nates from within our environment. It has to, because noth-ing can come in from outside except words" (197). That is to say, virtual reality is predicated on the distinction be-tween words and things, which is also the distinction be-tween inside and outside. Words can point outside—or come to us from outside—but they cannot actually take us there. They reveal the hollowness of the things that surround us and of the reality we inhabit, but they do not transport us to anyplace more solid. "Watching, wise physical ghosts from the full-life world" (213), such as Joe believes Glen Runciter to be, can leave behind traces—instructions for survival, inscriptions on coins, advertising jingles—in this our virtual world. But that is all; there can be no further contact.

Only Apparently Real. When Glen Runciter finds "Joe Chip money" in his pocket, at the end of *Ubik,* it is a sign that Runciter's world is also a "half-life" simulation. Con-trary to our previous impressions, we realize that the "full-life world" is as far beyond Runciter's grasp as it is be-yond Joe Chip's. Indeed, we may question whether such a "full-life world" even exists. All the signs suggest not. The consensus reality at the start of *Ubik* is a world in which, to use a phrase from Donna Haraway, "our machines are disturbingly lively, and we ourselves frighteningly inert" (152). The money, after all, has Walt Disney's picture on it. And the world is embroiled in a nightmarish conflict between "Psis" (telepaths and precogs), who invade other people's minds, and "Inertials," who nullify the Psis' powers. As Joe Chip explains, the Inertials are "a life-form preying on the Psis, and the Psis are life-forms that prey on the Norms" (26). In the midst of such a Darwinian life-and-death struggle, appearances are dangerously unstable. The world at the start of the novel is thus already a kind of virtual reality. As for the world that Dick himself lived in—and that we continue to inhabit, over a third of a century later—the criteria set forth in *Ubik* suggest that it too is nothing more than a simulation. For our world, too, is organized by money, ubiquitously commodified, penetrated by electronic media, and saturated with advertising messages, in just the ways that *Ubik* satirizes. No language is more familiar to us than that of the ads for Ubik that serve as epigraphs for each chapter of the novel: "Wake up to Ubik and be wild! Safe when taken as directed" (35). In our world, too, words are divorced from actions, separated from things and from bodies. Katherine Hayles suggests, in her reading of *Ubik*, that the barrier separating inside from outside, or the half-life world from the full-life world—a barrier that only words can break through—is a figure for Dick's authorial relationship to his text (188). This seems accurate enough to me, with the proviso that the author himself is also a fictional character, caught in an infinite regress. For, as Pamela Jackson points out, Dick himself came to see *Ubik* in this way. He felt that higher forces had somehow dictated the novel to him, and that Ubik itself (the aerosol spray can, equated with God) was the book's actual author. Dick identified with the figure of Joe Chip, and he tried to comprehend the reality of Orange County, California, on the model of Joe Chip's exploration of the half-life world; he insisted that, buried within advertisements and other base manifestations of popular culture, the trashier and more manipulative the better, there were true signs of redemption and divinity. Given all this, we are forced to conclude that Philip K. Dick is as fictional a personage as Glen Runciter or Joe Chip or (despite Dick's wishes to the contrary) God himself.

The Simulacral Body. In the contemporary imagination, virtual reality is often equated with disembodiment. A famous passage from William Gibson's *Neuromancer*, the canonical text of cyberpunk science fiction, describes how the book's protagonist, Case, "lived for the bodiless exultation of cyberspace." Case had "a certain relaxed contempt for the flesh. The body was meat." When his nervous system was hacked, so that he couldn't jack in to cyberspace any more, "it was the Fall." Case "fell into the prison of his own flesh" (6). In one sense, such hatred of the body is nothing new. It has a long history in Western culture, going back at least to the ancient Greeks. Plato says that we must leave our bodies behind if we are to ascend into the world of Ideas. More radically, and closer to Gibson's description of Case, the Gnostics explicitly equate fallen existence with being trapped in the flesh. Nonetheless, there is at least one crucial difference between the ancient and modern versions of "contempt for the flesh." It has to do with the way that the duality of mind and body gets lined up with that other duality of virtual and real. For the Greeks and Romans, mind or spirit is real, and materiality is the virtual term. Our world of matter and bodies is the realm of simulacra. For the Gnostics, the most extreme of the ancients in this regard, simulations are degraded and deceptive precisely because they are physical. But even the naturalistic schools of ancient philosophy define virtuality in physical terms. As Deleuze notes, Lucretius sees simulacra as "subtle, fluid, and tenuous elements" that "detach themselves from the surface of things (skins, tunics, or wrappings, envelopes or barks)" (1990, 273). These simulacra may be less solid than bodies, but they are still wholly material entities. In *Neuromancer*, to the contrary, matter and the body are regarded as actual, and virtual reality is seen (for good or ill) as an escape from the flesh. Gibson's novel describes cyberspace as "a consensual hallucination . . . ranged in the nonspace of the mind" (51). Today, we tend to think of simulacra as immaterial mental constructs, something that would have been unintelligible to the ancients. This change in the value of simulacra can probably be traced back to Descartes; his Evil Genius hypothesis marks the first occasion in Western thought that demonic deception is associated with the absence of matter instead of with its presence. Case is a strange sort of inverted Cartesian: he wants nothing more than to embrace the logic of the Evil Genius, to exult in pure hallucination, and to dismiss the physical world altogether. But of course, he is never able to do so. "The world, unfortunately, is real," as Borges (332) sardonically remarked.

The Visions Became Flesh. The virtual "half-life" world in *Ubik* is very much a physical one. One chapter describes, in excruciating detail, Joe Chip's overwhelming fatigue as he tries to climb a flight of stairs: "His world had assumed the attribute of pure mass. He perceived himself in one mode only: that of an object subjected to the pressure of weight" (173). A few pages later, we find the astonishingly visceral passage in which Joe is assaulted by Jory: "Snarling, Jory bit him. The great shovel teeth fastened deep into Joe's right hand. They hung on as, meanwhile, Jory raised his head, lifting Joe's hand with his jaw; Jory stared at him with unwinking eyes, snoring wetly as he tried to close his jaws. The teeth sank deeper and Joe felt the pain of it throughout him. He's eating me, he realized" (198–99). Dick thus discovers, at the heart of the "half-life" world, something like what the Stoics—as paraphrased by Deleuze—call the realm of passions-bodies and infernal mixtures: "everywhere poisonous mixtures seethe in the depth of the body; abominable necromancies, incests, and feedings are elaborated" (1990, 131). The virtual is realized or embodied, as it were, at a point of maximum physical density, a metastasis of impossibility, like a cancer cell or a cosmic black hole. At the opposite extreme from this horror, William Gibson imagines a light, lithe materialization of the virtual. Where *Neuromancer* was about the coming-to-consciousness of an artificial intelligence, his later novel *All Tomorrow's Parties* narrates the coming-to-body of an intelligent, virtual "emergent system." At the book's climax, the *idoru* (virtual pop star, or simulacrum) Rei Toei emerges out of cyberspace, and manifests herself in the material world. She appears, at once, in convenience stores all over the globe: a "naked Japanese girl," or more precisely a multitude of them, each smiling the same beguiling smile as she coolly steps past the amazed onlookers and into the chaos of the streets (268–69). Rei Toei's multiple incarnation is one of the great "nodal points in history," a privileged moment at which contingent circumstances converge to create a new constellation of forces. As a result of this convergence, "it's *all* going to change," history mutating in radically unforeseeable ways (4). Even if Gibson's whole scenario seems too much of an Orientalist fantasy, the crucial point is that binary code has become flesh. The virtual is realized or embodied, not violently, but in its most "subtle, fluid, and tenuous" form.

The Erotic Life of Machines. Chris Cunningham's video for Björk's song "All Is Full of Love" reimagines the android as a subtle and fluid being, as virtual as she is physical and mechanical. Björk appears as an android being assembled on a long platform, in a room of antiseptic white. Other machines are busy working on her. Their flexible arms poke and pry. They attach a panel here, tighten a bolt there. A cylinder turns, emitting a shower of sparks. A light flashes under an open hinge. Water gushes backwards, seeping out of the drain and leaping back into the spout. Nothing is inert. Everything has a cool, sensuous presence. Every mechanical object turns on its axis or glistens or thrusts and withdraws. Every material substance flows or splashes or spurts or sputters. We see these processes in extreme close-up. The video thus reveals the erotic life of machines. Why should Björk herself be any different? Soon, we see two Björk androids, instead of one. They face each other, singing by turns. A moment later, they are making love. We view them from a distance, in silhouette. They kiss and slowly caress each other's thighs and legs and buttocks. Their motions are so slow and stylized, as to suggest a superhuman state of grace. All the while, the other machines keep working on their bodies. Sexuality and reproduction are thus entirely separate processes, although they go on adjacently and simultaneously. Throughout, Björk's face remains blank and impassive. Her eyes, nose, and mouth are delicately modeled. Otherwise, the surface of her face is entirely smooth. Björk's eyes flutter, and her mouth moves slowly and precisely, as she sings of endless love. Her pronunciation is oddly toneless. Her voice is ethereal, almost disembodied. Shimmering washes of sound accompany these vocals. Densely layered strings lay down a thick, dissonant drone. Ghostly harp arpeggios rise out of the murk. A slow and steady synthesized beat grounds the song somewhat. But Björk's voice drifts away from the fixed pulse. She phrases the notes unevenly, now stretching them out and now shortening them. She hovers around the beat, without ever landing precisely on it. Time has become elastic. It has lost its forward thrust. It no longer moves at a fixed rate. It dilates and contracts irregularly, according to the modulations of Björk's voice. Visually, too, the video moves with a kind of suspended animation. There are no fast camera movements and no shock cuts or jump cuts. There are few colors to be seen. Nearly everything is a shade of white. The video's lighting ranges from a harsh white, to a muted blue-white glow, to a few white lines gleaming in the darkness. It's as if the world had been bleached and rarefied and chilled to nearly absolute zero. And in this digital coolness, all the pieces of the world have been recombined, following strange new rules of organization. They have congealed into new emotions and new forms of desire. In its own way, the machine is also a kind of flesh. Usually, we think of machines as being uniform in their motions. They are supposed to be more rigid than living beings, less open to change. But "All Is Full of Love" reverses this mythology. It suggests that robots might well be more sensitive than we are. They might have more exquisite perceptions than we have. They might respond, more delicately than we do, to subtler gradations of change. And they might have a higher tolerance for ambiguity. As Björk embraces Björk, in a quietly ecstatic feedback loop, the digital celebrates its nuptials with the organic.[48]

Birth of the Cool. "Cyborg" is defined by the WordNet online dictionary (Miller et al. 2001) as "a human being whose body has been taken over in whole or in part by electromechanical devices." The crucial phrase here is *taken over:* it implies, at the very least, a prosthetic replacement, if not an all-out viral invasion. I become a cyborg when some part of my actual body is taken over by the virtual. My sensory apparatuses, and my organs, are always being replaced or extended by technological devices. This process is coextensive with the whole of human culture. As McLuhan says, every medium, which is to say every technology, is an "extension of ourselves" (1994, 7). But in each instance of technological change, he adds, we misrecognize the very extensions that we have created and see them as forces alien to ourselves. As a result, "man is impelled to extend various parts of his body by a kind of autoamputation" (42). The first stone tools and the first spoken language already worked this way. I extend the power of my hand or my mouth or my brain only at the price of excising the original organ—whether literally or figuratively—to make room for its replacement. Each time we extend ourselves technologically, some part of the real gives way to the virtual. This is why every cultural innovation is attended by an ambivalent sense of loss. And this is also why we tend to equate *virtual* with *disembodied,* even though it would be more accurate to use it as an equivalent for *prosthetic.* In a certain sense, then, we have always been cyborgs, even if, by the strictest meaning of the term, the transformation only happened recently, when our prostheses became "electromechanical devices." Perhaps just wearing glasses doesn't quite make me a cyborg, but watching television certainly does. It is in this context, I think, that we can best appreciate the utopian prospect of the cyborg, as imagined by Donna Haraway. The cyborg is a mythical being who cuts across all three of the "crucial boundary breakdowns" that characterize postmodern existence. There are now only "leaky distinctions" between the human and the animal, between the biological organism and the machine, and between the physical and the nonphysical. These distinctions have not been altogether eliminated, but the boundaries that used to define them have become "permeable." The cyborg is the very figure of this permeability. It is fully physical, but light and cool: a "subtle, fluid and tenuous" form of materiality. In their indifference to binary categorizations, their easy trafficking between real and virtual, Haraway writes, "cyborgs are ether, quintessence" (151–53).

Dancing with the Virtual Dervish. Diane Gromala's virtual reality installation *Dancing with the Virtual Dervish*[49] cuts across the leaky boundary between virtual and real. This is how it works: I slip on a head-mounted display and clutch a pointer in my right hand. Immediately I find myself in darkness, a night sky with distant stars. Electronic music plays faintly in the background. I turn about slowly, trying to get my bearings. I shake my head left and right and up and down. By stretching out my hand and pointing, I am able to fly through space. By pulling my hand back toward my body, I can slow down. I start and stop awkwardly, abruptly. I realize that it will take a while for me to orient myself and to learn how to navigate through this strange new world. Nonetheless, not all is confusion. After a while, I notice a bright line cutting through the gloom. I follow this line the best I can, and finally something substantial comes into sight: a strange three-dimensional object, a bizarre architecture of vaults and arches and tunnels and jutting projections. As I approach the enigmatic object floating in the void before me, I am able to make it out more clearly. Now I can see that it is the inside of a human body, a skeleton and torso viewed from beneath the skin. The main series of arches is a rib cage. The other, smaller branches are limbs. The thick, short tubes are veins and arteries. I swoop down the spine and in and out among the ribs. Sometimes I just enjoy the sense of movement. At other times, I stop to look around. These twisted spaces are richly textured in an alluring combination of density and plasticity. I can easily recognize some shapes. Others are like objects seen in dreams: they seem richly suggestive, but I'm unable to figure out precisely what they are. Many of the surfaces are inscribed with words: mysterious and fragmentary texts expressing erotic yearning and loss. From time to time, also, I glimpse another human figure in the space. He is a dancer, wearing a virtual reality helmet just like me and turning and twisting his body in fluid, abstract movements. I can never quite get near him; he is just a small image floating in a rectangular membrane. For a moment he comes into view, then I lose him again. There are also small solid objects lodged among the bones and blood vessels. After a while, I realize that they are the body's internal organs. There's the liver, and there is one of the kidneys. That one just below me must be the heart, for I hear its faint, yet steady, beating as I approach. When I enter one of these organs, the display blacks out for a moment. Then a completely different landscape opens out before me. Now I'm immersed in a space for which I have no points of reference. Blue slats stretch out in all directions and at all angles. The piece has a number of these sites enfolded within the organs: secret spaces, worlds within worlds.

Inside Out. The landscapes of *Dancing with the Virtual Dervish* are organic and abstract all at once. They have little to do with the geometric matrices and photorealistic textures that make up so many computer-generated, virtual spaces. Rather, they are topologically complex, folded and curved, with very few straight lines. Although they are simulations, generated by a machine, they are not transparent to thought. They refuse easy legibility and resist our desire to reduce them to shapes with which we are already familiar. And there is a good reason for this. The main virtual space of *Dancing with the Virtual Dervish* is in fact a rendering of the insides of Gromala's own body. She based the piece upon actual MRI scans of her viscera. The scans were taken when she was suffering from chronic pain, which her doctors were unable to explain. The doctors wanted to capture her inner, bodily experience and put it into a form that they could easily understand. They deployed the most advanced tools of medical technology in order to objectify, and thereby validate, her pain. But by taking the scans and transforming them into art, Gromala reverses this process. She re-subjectifies her body, reclaiming these high-tech medical images as her own. By moving inward, lingering over the landscapes of her body, and becoming so acutely aware of these visceral spaces and densities, Gromala turns technology back into flesh and transforms her pain into pleasure. *Dancing with the Virtual Dervish* is therefore a strangely enigmatic piece: at once alien and intimate, public and private, hermetic and expansive. It brings us close to the core of Gromala's pain, yet it doesn't actually allow us to feel this pain or share it. In a certain sense, it places us at a vast distance from the phenomenal feeling of pain, for the overall feel of being immersed in the piece is more calm and contemplative than anything else. Except that this is a very different sort of distance than the one created by the original MRI scans. For in *Dancing with the Virtual Dervish,* distance itself is made intimate, enfolded within Gromala's own body. And to feel distance so intimately, so viscerally, is perhaps part of what it means to be in pain. *Impersonal intimacy* might seem to be an oxymoron, but in Gromala's work, this is one sense of what it means to be virtual.

Gravity. When I explored *Dancing with the Virtual Dervish* in 1999 at the University of Washington's Human Interface Technology Lab, I faced the problem of insufficient computing power. The 3-D rendering took too much time to be altogether convincing. I couldn't fly through virtual space as fast as I would have liked. The transitions, when I entered one or another of Gromala's internal organs, were abrupt and discontinuous; they should have been more like cinematic dissolves. I had to hold a mere pointer in my hand, instead of being able to wear a dataglove, let alone one of those full-body VR suits that we are promised sometime in the future. All this reminded me that I was just in a simulation. But even more powerful computing equipment would not have made the experience totally seamless. For the room I was in retained its hold on my peripheral awareness. I could sense its light just past the edges of my goggles. I felt the weight of the VR headset as it sat on my skull. I felt the force of gravity holding me down, even as I was flying through virtual space. There was a cord connecting my helmet to the computer, and sometimes it would get all tangled up; I had to be careful to avoid tripping over it. As I made motions with my hands and hips, I became self-conscious, aware of how absurd I must have looked from the outside. I have seen other people using VR gear; it's always a bit disturbing to see them flailing about in empty space without any context for their actions. Worst of all, after being inside the piece for half an hour or so, I started to experience a mild vertigo. This is the well-known *simulation sickness:* the incoherence between the virtual sensory input one receives and the sensations that one continues to get from one's surroundings becomes too much for the body to handle. The proprioceptive system breaks down in confusion. Virtual reality enthusiasts tend to argue that these are merely transitional problems, to be overcome once the technology is more fully developed. But Gromala does not try to minimize such interference effects; rather, she explicitly builds them into the very structure of the piece. *Dancing with the Virtual Dervish* turns upon the complex entanglements of weight and weightlessness, fluidity and congealment, place and placelessness, incarnation and disembodiment; in short, upon a whole series of confused encounters between the actual and the virtual. Gromala suggests that these confusions are not contingent and eliminable side effects, but intrinsic to the way that virtual reality works.

Simulations are effective because they address our embodied senses and not because they would somehow allow us to escape from the body. There is always an ambiguous overlap between what is real and what is not, between the virtual environment that engulfs my senses and the actual surroundings that I cannot altogether escape. Gromala's exploration of her own insides is less a simulation of bodily experience than it is an intense embodiment of simulation itself. *Dancing with the Virtual Dervish* acknowledges the simultaneous coexistence, and the mutually interpenetrating influence, of virtual and physical worlds, as well as exposing the virtual world as something that is itself material and physical.[50]

The Wedge. In K. W. Jeter's *Noir,* the virtual is a special place: a nocturnal zone known as the Wedge. Sometimes the Wedge seems to overlay the real world, a spectral double of its streets and buildings and private spaces. Other times, it seems to lie in an actual spatial location, like the red-light district of an older city. In either case, the Wedge is a murky realm of hidden, shameful, and disavowed pleasures. It's a place where kinky fantasies come to a shadowy, anonymous half-life. Bodies are sculpted into fantastic shapes by a bizarre variety of prostheses and amputations. Everyone's skin is covered with black-ink tattoos, and these tattoos are alive: they swarm over the flesh "like decorative koi or human-eyed piranha" (7), they morph and mutate, they migrate from one body to another. Few human beings actually enter the Wedge themselves. Instead, they send proxies, quasi-human androids known as prowlers, who sample the fruits of transgression and bring the bittersweet taste back to their owners. It is literally a matter of the sense of taste. The prowlers transmit their experiences by means of a deep-tongued kiss that leaves a burning sensation behind, "the hot copper taste of coded flesh" (3). This is how the digital stamps its mark upon the analog. When a prowler kisses me, what I receive from him or her is the memory of a violent sexual encounter: "all the pleasures of remembering and none of the risks" (357–58). I do not actually experience these blows, thrusts, and caresses; they do not affect me in the present. But now I remember them, as vividly as anything that ever did happen to me, "the engorged memories popping out from each other like an infinite series of Chinese boxes" (355). In this way, the virtual world of the Wedge is a place of palpable nostalgia. It's as real as anything not immediately present could ever possibly be: as real as the past, as real as words, as real as my sense of myself.

Now or Never. In *Noir*, the virtual is the realm of memory, and the actual is what takes place in the immediate present. But of course, this means that the virtual is nearly everything. The living present is only the merest sliver of an instant, "a sliced-apart microsecond" (339) wedged between the vast expanses of past and future. By the time I am able to take cognizance of such a moment, it is already over and gone. The intensity of the present is too great for consciousness to register or language to contain. It is literally unbearable, which is why I must use a prowler to experience it for me. For even the vicarious sting of the prowler's kiss is overwhelming: a "rush of sensation and memory data" (339), a tongue "locked in bandwidth rapture" (4), a body penetrated and crucified by "the feedback of flesh distant in time and space" (5). The prowlers themselves are cool, digital beings, creatures of "latex and soft industrial resins" (338), whose synthetic flesh carries a chill "ten degrees lower than [our] own body temperature" (6). They are oddly anonymous, like extras in old movies: "you see them and you don't see them—the extras, I mean. They exist in that world, they're even necessary—but you don't remember them...there's nothing in their faces to snag onto normal peoples' memories" (335). Yet the blanker and more forgettable these prowlers are in themselves, the more singular and powerful the memories they bring back for my consumption. In the world of *Noir,* prowlers are too expensive for most people to afford. They are only used by a privileged—or depraved—few. Nonetheless, they have a significant place in the book: they provide a model of subjectivity and attest to the commodification of lived experience. An actual self can only exist if it is coupled with a virtual, anonymous prosthesis; an immediate present can only be experienced if it is grounded in the pastness of memory; and an analog interface to the world can only function if it is patched into a network of digital processors. In all these ways, the prowler is what Derrida would call the *supplement* of the subject: something that is external to its nature and yet necessary for it to function at all. Perhaps this is why the prowler has a certain stigma attached to it. Its very existence is somehow shameful, for it testifies to my own insufficiency. To have dealings with a prowler is to feel "something between fear and disgust"; although, Jeter adds, this is an ambiguous state "in which those terms no longer ha[ve] a negative connotation" (317).

Supplemental. "We possess art," says Nietzsche (1968b, 435), "lest we perish of the truth." In a similar way, we have virtual reality, lest we perish of finding the world to be incomplete. Virtuality is both the cause of the world's incompleteness and the remedy for it. The logic of virtuality shares in the dizzying convolutions of Kant's antinomies of reason and Derrida's supplementarity. Both Kant and Derrida seek to track down metaphysical illusions that are nevertheless inherent to the very structure of our rationality. That is to say, they seek to point out, from within, the ways in which we are, as Nietzsche says, *"necessitated to error"*

(1968a, 37). Such a logic runs through any attempt to circumscribe the virtual. *On the one hand,* the world seems phony, or unreal, precisely because so much of it is virtual. Things have been hollowed out, reduced to mere facades, replaced by simulacra of themselves. This is the America of shopping malls and theme parks and media spectacles, so powerfully described in the writings of Philip K. Dick and Jean Baudrillard. The problem with virtuality, from this point of view, is that it is both too little and too much. It is flimsy and superficial, devoid of real depth or serious content. But this very deficiency leads to its alarming capacity for viral proliferation. So the virtual is also excessive: it transforms the whole world into its own dazzling image, a seamless web of hyperreality. *On the other hand,* and at the same time, virtuality is the one saving grace that makes up for the world's otherwise chronic unreality. It is a heightening enrichment of experience, whether in McLuhan's sense of media as the "extensions of man" or in J. G. Ballard's vision of the new forms of desire that arise from our involvement with new technologies.[51] Here, virtuality seems to be at once a vital necessity and a superfluous luxury. To manufacture the virtual is a serious task. It is the labor of transforming the real, whether this be seen (in the Hegelian manner) as progress toward some definite goal, or (in accordance with more recent fashion) as the ongoing adaptive behavior of a complex, self-organizing system. But there is always something left over from such labor, something thrown out or abandoned as an inessential by-product. Virtuality is also this remainder. It is the frivolous enhancement of the senses, the useless adornment of the actual, the barely, only marginally real: in short, what Bataille calls *unproductive expenditure.* Excess and deficit, useful labor and waste, amputation of the body and enhancement to it: the virtual is all of these things at once. Virtuality cannot be circumscribed or properly defined because its essence is to be inessential; it does not have a "real," or proper, place.

Virtual Light. The actual and the virtual are mutually dependent. Neither is meaningful without the other. Every empirical object has its aura of virtuality; every virtual state is grounded in some sort of materiality. The virtual cannot be opposed to the actual in the way that the soul is traditionally opposed to the body. It is better to say, paraphrasing Kant, that the virtual without the actual is empty, while the actual without the virtual is blind.[52] The virtual illuminates the actual, but it is nothing without the actual's support. The relation between actual and virtual is something like the one between hardware and software. A computer is able to calculate, and thereby to simulate, an indefinite number of possible worlds. But this can only happen because each of these worlds is strictly correlated with a particular physical state of the machine. Now, these machine states are themselves entirely actual, while the possible worlds that they support are virtual. These two dimensions are coextensive, yet entirely different in nature. Small changes in the actual physical state of a system may correspond to widely different virtual events and even to entirely different virtual worlds. This is why a software program can run, with identical results, on many different pieces of hardware, and why, conversely, a single piece of hardware can run many different sorts of software. Arguing from this disjunction, futurists like Ray Kurzweil foresee the possibility of "downloading your mind to your personal computer" (2000, 124). It's a question of learning how to copy the contents of the mind in sufficient detail and then installing that copy in a machine other than the brain: "we don't need to understand all of it; we need only to literally copy it, connection by connection, synapse by

synapse, neurotransmitter by neurotransmitter" (125). Kurzweil seems to believe that we can do this without worrying
about the underlying hardware of the brain; we can just
ignore "much of a neuron's elaborate structure," he suggests, since it only "exists to support its own structural integrity and life processes and does not directly contribute
to its handling of information" (125). But contra Kurzweil,
this distinction is entirely bogus, for the brain's "handling
of information" is itself a "life process" that depends upon,
and in turn affects, the "structural integrity" of the neurons. Indeed, one could not "literally copy it" in the first
place unless one paid attention to the "elaborate structure"
underlying it all. Kurzweil does entertain the idea that a
downloaded mind will need some sort of new body, if only
because "a disembodied mind will quickly get depressed"
(134). But he fails to grasp the full extent of the reciprocal
correlation between the mind and the body, or software
and hardware, or the virtual and the actual.

Preemptive Multitasking. Fantasies of downloading
the mind into a computer seem to depend upon an overly
simplistic extrapolation from the idea that the same sequence of code can run on many different machines. In
fact, this scenario of mind transfer is about as plausible as
the claim, made in a television commercial for IBM in 2001,
that you can "download an entire warehouse" over the
Internet. Such operations will only be possible when nano-
technology allows for the "nanofaxing," or precise replication over distance, of physical objects, as in Gibson's *All
Tomorrow's Parties* (268–69). To duplicate my consciousness,
or transfer it to another location, I will need to know how
to reproduce the biological hardware of my brain and spinal
cord, as well as the mental software containing my memories, sensations, emotions, and thoughts. In the meantime,
it might be best to forget about running the same software

on many pieces of hardware and focus instead upon the converse idea of running multiple software programs at once on a single piece of hardware. Why worry about transporting myself into another body when I still haven't realized how many different selves are present in this body that I already have? People with Multiple Personality Syndrome are said to display different physiological patterns—different voices, different pulse rates, even different levels of cholesterol—depending on which personality is in the forefront at a given moment. Thus the same actual body can support many different virtual selves. I argue elsewhere (1997, 147–56) that the phenomenon of multiple personalities, along with Pierre Klossowski's notion of demonic possession, gives us a better paradigm for postmodern subjectivity than anything we can get from psychoanalysis. You cannot be one without being at least two. We are all at least potentially multiple, even if most of us do not suffer from the oppressive consciousness of being so. In recent years, increasing numbers of multiple "households" themselves have come to reject the idea that their multiplicity should be regarded as a medical "disorder." In particular, they resist the received psychiatric dogma that the experience of multiple personality syndrome either can or should be "cured" by integrating all the personalities into a single self.[53] More is lost than gained from such a normalizing reduction. The point is not to eliminate these multiple identities, but rather to get them to talk to one another and to find ways for them to continue cohabiting with each other, in their one shared body, without too much distress or conflict.

The Turing Test. Of course, computer multitasking isn't really simultaneous at all; or better, its simultaneity is an optical-conceptual illusion. A digital computer, like any other Turing Machine, is a serial device that only processes a single instruction at a time. What actually happens during multitasking is that the processor continually switches between the different programs that it is running. And if this switching takes place rapidly enough, compared to the rate at which our own minds work, then the effect upon us is one of simultaneity. That is to say, a computer's operations are *actually* linear and sequential, but *virtually* they are multiple and synchronous. Now, this is precisely the opposite of what happens in human brains. Our underlying neural activity seems to be massively parallel. But nearly all of this activity is unconscious. Human consciousness, on the other hand, is experienced serially. Language, too, is a linear phenomenon, though its underlying structure is synchronic. Nicholas Humphrey suggests, therefore, that consciousness is something like a serial interface to the underlying parallel processing that fundamentally constitutes mental activity. Consciousness is not thought itself, but an easily accessible simulation of thought. The brain's operations are *actually* multiple and synchronous, but *virtually,* they are linear and sequential. Or to put the same point in slightly different terms: the brain is not a Turing Machine. It does not work by means of algorithmic calculations. Rather, it operates according to entirely different principles, ones that we do not currently understand. Nonetheless, the calculations of digital computing—and consciousness—can often arrive at the same conclusions as the unconscious brain; that is to say, many brain processes can at least be simulated by Turing Machines.[54] Even brain parallelism can be simulated to some extent through neural nets or multiple processors. The question still remains, though, as to whether or not all the brain's operations can be simulated by Turing Machines such as digital computers and conscious minds. This is the real issue at stake in current arguments about the powers and limitations of artificial intelligence, although it is often overlooked by advocates on both sides of the debate. Partisans of "strong AI," such as Daniel Dennett (1998), seem to assume, without evidence, not only that algorithmic computation can simulate brain activity, but that the brain actually operates this way. Opponents of "strong AI," such as Roger Penrose and John Searle, rightly reject this assumption, but then they make the equally dubious claim that consciousness is a uniquely privileged aspect of mental activity that cannot be produced by computational means. Against both camps, I want to suggest two things. First, that the question of whether the brain actually is a Turing Machine and the question of whether the brain can be simulated by a Turing Machine are entirely separate issues. Second, if there does turn out to be some sort of brain activity that cannot be simulated by a computer, then this activity will turn out to be unconscious as well, and indeed radically inaccessible to consciousness.

Deus Ex Machina. Let's assume, just for the sake of argument, that artificial intelligence is an actual possibility. Let's even assume that its development is inevitable, once we have electronic media at all. Artificial intelligence, we might say, is an unplanned, emergent property of any sufficiently dense system of connections. When the network expands beyond a certain critical threshold, it sets in motion a new logic of its own. Its nodes become spontaneously self-aware. Intelligence may not inhere in any single physical device, but it extends, distributed, all across the Web. What might the nature of such a machine mind be? Usually, artificial intelligence is seen as being radically transcendent, as in Kevin Kelly's creepy vision of a "hive mind," or Pierre Teilhard de Chardin's idealist projection of the teleologically evolving noosphere. From a more strictly technological perspective, Hans Moravec (2000) traces a progression "from mere machine to transcendent mind." He gleefully foresees a future in which our robot "mind children" (Moravec 1990) inherit the earth, making humanity

itself obsolete. Ray Kurzweil (2000) is somewhat more re-
strained than Moravec since at least he envisions the fusion of organism and machine, rather than the total displacement of the former by the latter. But Kurzweil still
sees "machine-based intelligences" as possessing well-
nigh magical powers. They will "not [be] tied to a specific
computational-processing unit (that is, piece of hardware)";
instead, they will be "able to manifest bodies at will: one or
more virtual bodies at different levels of virtual reality and
nanoengineered physical bodies using instantly reconfig-
urable nanobot swarms" (234). More interestingly, William
Gibson's *Count Zero* presents its autonomous AIs as aes-
thetically inclined, sublimely detached observer-deities of
cyberspace, who assume the personae of the loas of Haitian Voudoun mythology. The problem with all these accounts of artificial intelligence is that they massively over-
rate the power and value of "mind." They imagine that
thought, or more precisely computation, is a force of infinite
potential, without limits. Once intelligence has been freed
from its fleshly encumbrances, it is paradoxically able to
penetrate and manipulate matter and master the physical
world with almost no resistance. AIs will be like gods; or
even better, in Kurzweil's amazing phrase, they will be "all
basically entrepreneurs" (243), purveyors of an endlessly
expanding, frictionless celestial capitalism.

Posthuman Politics. The real problems with artificial
intelligence are social, political, and economic. In his "Fall
Revolution" series, Ken MacLeod considers the politics of
AI, as it is played out in different social and economic contexts. Consider Dee Model, a "gynoid" (38) or "human-
equivalent machine" (13) from *The Stone Canal.* Her body is
human, but her mind is not. Rather, she has a machine-
based intellect, "vast and cool and unsympathetic" (7), that
is divided among various autonomous programs and sub-

routines. Her sense of self, for instance, is not the core of her being, but just the output of a particular software program called Self that sits alongside such other programs as Spy, Soldier, Secretary, Sex, and Story (9 and passim). The Jargon File, an online lexicon of hacker slang,[55] wonderfully defines *program* as "an exercise in experimental epistemology"; Dee is a perfect embodiment of this definition. "Technically and legally I'm a decerebrate clone manipulated by a computer," she explains at one point. "Neither component is anything but an object, but I feel like I'm a person" (13). Comparing Dee to a human being with an organically grown brain, we could say that the same result—a sense of personhood—is achieved by radically different means. In other words, Dee is a simulacrum, in precisely the sense defined by Deleuze (1990): "it still produces an effect of resemblance; but this is . . . produced by totally different means than those at work within the model" (258). But does it matter that Dee's Self software only *simulates* self-consciousness? MacLeod suggests not; much of *The Stone Canal* narrates Dee's successful quest to be recognized as a full legal and economic subject. MacLeod makes a similar point in *The Cassini Division.* For most of the book, Ellen Ngwethu scorns the idea that artificial intelligences might be conscious: "they may give the appearance of sentience, but if they do, it'll be a protective coloration . . . they can no more feel than the eye-spots on a butterfly's wing can see" (87). But Ellen is forced to change her opinion after she nearly dies, and her own mind has to be restored from a software backup: "So now I knew," she says. "I knew how a simulated mind experienced the world. There was no difference whatsoever" (228). When it comes to sentience, the means are irrelevant; only the result is important. Thus MacLeod endorses the "strong AI" position. But he also concludes—in opposition to most AI enthusiasts—that these AIs will not be transcendent. Just like human beings, they will be contingent and limited entities, situated in a particular time and place, and will have their own particular needs, interests, and values. Machine intelligence might well be qualitatively different from ours, as well as quantitatively deeper and faster, yet ultimately, the differences between machine and human perspectives are cultural and political matters rather than metaphysical ones. They are not necessarily any greater than the cultural differences that already exist among human communities, like those among the socialists, the free-market libertarians, and the Greens, that are played out over the course of MacLeod's four volumes.

The Singularity. The mathematician and science fiction writer Vernor Vinge seems to have been the first to have announced the coming Singularity: the moment at which the creation of a superhuman, machine-based intelligence will result in "a regime as radically different from our human past as we humans are from the lower animals . . . From the human point of view this change will be a throwing away of all the previous rules, perhaps in the blink of an eye, an exponential runaway beyond any hope of control." More recently, Ray Kurzweil (2001) has amplified Vinge's prophecy: "The Singularity is technological change so rapid and so profound that it represents a rupture in the fabric of human history." Vinge believes we will reach this point by 2030 at the latest. A bit more conservatively, Kurzweil calculates that we might have to wait until 2049 or so. Using psychedelic insights and Mayan cosmology, instead of extrapolating from Moore's Law, the late Terence McKenna (1994) came up with the somewhat earlier date of 12 December 2012 for the end of history.[56] No matter when it happens, however, the Singularity marks a point of no return for the human race—or at least for that small portion of it that is rich and well-connected enough to get access to all the cool technology. They will get to be as gods (or entrepreneurs). Everyone else will simply be cast aside; Kurzweil ominously warns that, in the aftermath of this transformation, "humans who do not utilize [neural] implants [will be] unable to meaningfully participate in dialogues with those who do" (234). Maybe, just maybe, if we are lucky—and also depending upon just how they define meaningful dialogue—the superhumans will choose to treat us as kindly as we today treat cats and dogs. Or maybe not. In any case, Kurzweil and Vinge seem blithely indifferent to the fate of the unwired billions. Artificial intelligence is the next step in evolution, they seem to be saying, and our only viable option is to Get With The Program.

I'll Teach You Differences. Kurzweil (2001) defines the word *singularity* as follows: "In mathematics, the term implies infinity, the explosion of value that occurs when dividing a constant by a number that gets closer and closer to zero. In physics, similarly, a singularity denotes an event or location of infinite power. At the center of a black hole, matter is so dense that its gravity is infinite. As nearby matter and energy are drawn into the black hole, an event horizon separates the region from the rest of the Universe. It constitutes a rupture in the fabric of space and time." One might add that a singularity is also a point in the graph of a function for which there is no slope or derivative; a point of phase transition in matter, like the boiling and freezing points of water; and a bifurcation point in chaos theory. A mutation in biology might even be thought of as a singularity—understanding that, as Dawkins says, the effect of a particular gene can only be understood negatively and differentially, in terms of the change that it makes, "*all other things being equal*," when it is substituted for any other allele found at the same location in the chromosome (1989, 37). For Deleuze, "singularities are turning points and points of inflection; bottlenecks, knots, foyers, and centers; points of fusion, condensation, and boiling; points of tears and joy, sickness and health, hope and anxiety, 'sensitive' points"; the main thing being that, in all these instances, the singularity "is opposed to the ordinary" (1990, 52). The singularity is a point of discontinuity, a point where linear extrapolations break down. In situations of "deterministic chaos," the smallest cause can have the most far-reaching effects. Variations in starting conditions too tiny to measure can lead to startlingly different outcomes. Even something as familiar as the process of water freezing and becoming ice is mysteriously transformative in this way: we would never expect it if not for the fact that, precisely, it is so well-known to us from everyday experience. (Think of the passages in García Márquez's *One Hundred Years of Solitude* about the singular discovery of ice.) Another way to put all this is to say, with Deleuze, that every singularity is an event; the play of singularities constitutes a *history* of a particular sort (52–53). Deleuze, like Foucault, stands for a discontinuous, contingent, and continually creative counterhistory, in opposition to the traditional notion of history as continuity and dialectical progression (or regression).

On the Paralogisms of Virtual Reason. Nietzsche, at least, presents his Superman as a being who is radically contingent. The *Übermensch* stands at the far end of an abyss that we ourselves cannot cross. He is someone or something that we will never grasp. In their invocations of the superhuman, Vinge and Kurzweil are nowhere near so modest. The problem with their speculations about the Singularity is not that they see the future as being different from, and radically discontinuous with, the past. The problem is rather that they translate this difference into something like a necessary and continual upward progression to a state of transcendence. According to Kurzweil, not only is technology growing at an exponential rate, but even "the rate of exponential growth is itself growing exponentially" (2001). He proves this with numerous graphs, all of which feature lines that curve relentlessly upward. In circumstances of such radical change, Kurzweil admits, "some would say that we cannot comprehend the Singularity, at least with our current level of understanding, and that it is impossible, therefore, to look past its 'event horizon' and make sense of what lies beyond." But Kurzweil doesn't let this objection bother him. For he wants to have it both ways; he announces that the future is unimaginable, and in the very same breath goes ahead and claims to accurately imagine it. Kurzweil's reasoning is based on what Kant called a "transcendental paralogism": a syllogism that is formally incorrect, because it claims to deduce positive concepts from a proposition that in itself is only a logical form devoid of any particular content (1996, 382). The target of Kant's critique was the "rational psychology" of his day, which presumed to deduce the immortality of the soul from the mere propositional form of the *cogito,* or what Kant calls "the simple, and by itself quite empty, presentation I, of which we cannot even say that it is a concept, but only that it is a mere consciousness, accompanying all concepts" (385). In a similar way, Kurzweil claims to deduce empirical consequences—the cybernetic immortality of the soul—from the mere form of technological change. He sees historical discontinuity itself as the final term of a rational progression.

The Rapture for Nerds. Ken MacLeod, in sharp contrast to Vinge and Kurzweil, sees the Singularity in the context of ongoing social and political conflict. One of his characters sarcastically calls it "the Rapture for nerds!" (1999, 90). This remark targets both the elitist exclusivity of the Singularity and the apocalyptic fervor with which it is greeted by the chosen few. In MacLeod's account, history doesn't end with the Singularity, it just takes on certain additional players and develops a few new twists. As Ellen Ngwethu subsequently recalls it, speaking with no little exasperation: "the Outwarders—people like ourselves, who scant years earlier had been arguing politics with us in the sweaty confines of primitive space habitats—had become very much not like us: post-human, and superhuman. Men Like Gods, like" (11). What can you do, she seems to be asking, with people like that? In MacLeod's series of novels, this question gets at least two answers. On one hand, AI technology is integrated into the texture of everyday life, which is thus enhanced, but otherwise goes on much as before. This is especially the case on the capitalist world of New Mars in *The Stone Canal*. Besides Dee Model, there are all sorts of AIs, some of them with citizenship rights, others defined as property and worked as slaves, yet others embedded in larger devices, and still others that function autonomously, outside of human society, as a sort of spontaneously evolving mechanical fauna. On the other hand, there are the AIs who opt for the Vinge/Kurzweil program of "running [the mind] a thousand times faster, and expanding its capabilities...just by *plugging more stuff in*" (1999, 97). These superhumans move to Jupiter to get away from the rest of humanity and form their own society. Their frequently hostile transactions with the people of the Solar Union, and their eventual extermination, are narrated in *The Cassini Division*. We learn that they have an extensive history, including episodes of imperialist aggression, a period of madness and near catastrophe, and "the repeated rise and fall of post-human cultures in virtual realities" (83). We also learn that they have not become any sort of "hive mind" or "collective entity" (194); despite sharing memories and information, they remain separate individuals, who argue and negotiate and make political decisions (137). And these erstwhile nerds are also extraordinarily beautiful: "deep violet eyes in gigantic faces, faces sweet and calm as those of any imagined angels. Their bodies too were like angels': with long, trailing cascades of gold or silver or copper hair, and sweeping diaphanous robes of rainbow light, each breastplated with a sunburst of jewelled filigree," and with "beating wings" that are "perfect parabolas, curved like magnetic fields, shimmering like polar lights; wings made from aurorae" (116).

The Lady or the Tiger. If it isn't political, then most likely it's just a fashion statement. Should we take the beauty of the posthumans on Jupiter at face value? Or should we, rather, endorse the suspicions of Ellen Ngwethu? With her cynicism fueled by a knowledge of evolutionary biology, she sees this beauty as a "mind virus" or a "killer meme": "a lure, precisely calculated to trigger our aesthetic reactions" (MacLeod 1999, 118). Of course, these alternatives are not mutually exclusive. MacLeod's AIs are clearly playing for high political stakes—they have none of the detached generosity of Gibson's cyberspace loas, for instance—but this doesn't nullify the beauty that they create. Our awareness that a certain gesture has been "calculated to trigger our aesthetic reactions" does not prevent our reactions from being triggered. We can override our aesthetic responses, as Ellen does, but we cannot stop ourselves from having them. In this way, the sense of beauty is different from other feelings. I am not likely to remain sympathetic to someone, for instance, if I come to realize that they have cold-bloodedly manipulated me in order to arouse my sympathy. But the posthumans of Jupiter still seem beautiful to me, even after I learn that they have crafted this beauty just in order to seduce me. The sense of beauty persists, because it is, as Kant says, disinterested. It's a feeling that *"cannot be other than subjective"* (1987, 44), for it is indifferent to the existence and the qualities of the object to which it refers. But at the same time, this feeling of beauty is strangely impersonal; it remains entirely independent of my needs, interests, and desires (51–53). And that is why beauty can be so dangerous. It can be used, as the post-humans on Jupiter seem to be using it, to lure us to our own destruction.

Foglets. Ray Kurzweil puts great stock in J. Storrs Hall's notion of foglets. These are intelligent, tiny robots—or nanobots—each about the size of a human cell. As foglets float through the air, they "grasp one another to form larger structures" and "merge their computational capacities with each other to create a distributed intelligence" (Kurzweil 2000, 145). In this way, the foglets join together in swarms, producing what Hall calls a *Utility Fog*. Human beings can also download their minds into these foglet swarms. The result, according to Kurzweil, is that "physical reality becomes a lot like virtual reality." That is to say, the world becomes infinitely, instantaneously malleable. A Utility Fog can rearrange its surroundings into any desired configuration, although, as Kurzweil admits, "it's not entirely clear who is doing the desiring." For instance, if you are hungry, foglet technology can "instantly create whatever meal you desire" (145). The only problem with this scenario is that, if you are a swarm of foglets, you will never be hungry for human food in the first place. You won't have a mouth, or a tongue, or taste buds, or a stomach. Of course, you will be able to simulate the pleasurable sensation of tasting gourmet food, but if you can do that, then why even bother creating an actual meal? All this is to suggest that, despite its name, the Utility Fog does not do much of anything useful. Like so many virtual and AI technologies, it is not a productivity tool, but a means of autonomous pleasure, like a high-tech vibrator or dildo. Such is clearly the case in *Transmetropolitan* (Ellis and Robertson 1998b, 72–92). As one Fog

Person explains, "we have no physical needs. All we have to do is amuse ourselves. Being regular humans can get in the way of that" (85). Mostly, it seems, the Fog People are virtual aesthetes and dandies, who pass the time by having impalpable sex (91) and by fabricating whimsical objects out of "air and dirt and whatever else is around" (83). Nothing could be further from Kurzweil's vision of AIs as hardworking, hard-computing entrepreneurs.

Wetware. One character in *Transmetropolitan* can't understand why anyone would ever want to be transformed into a foglet swarm: "If you're bored of your body, you could buy a new one, or temp, or even go transient. Why become dust?" (84). The answer, perhaps, is that whoever does this is "queer for machinery" (77). It's just one particular kink among many others. But it's one that AI enthusiasts, like Kurzweil and Moravec, share. They love to talk about DNA code, evolutionary algorithms (Kurzweil 2000, 294–97) and the like; but they tend to get squeamish when it comes to the actual messiness of biological "wetware," or anything else that isn't dry as dust and silicon chips. It's a long way from the orderly, progressive computations of classic AIs like Kurzweil's "spiritual machines" (or even of Kubrick's HAL, for that matter) to the delirium of "splices" (genetically hybridized organisms), "tropes" (designer neurochemicals), and "mods" (biological implants), as described in Paul Di Filippo's short story collection, *Ribofunk*. The merger of biology and information processing works in both directions. The genome is becoming ever more reduced to algorithmic, digital calculations, but thought and calculation are themselves becoming ever more subject to the vagaries of analog, phenotypic embodiment. If "the biologic bank is open," as Burroughs says, "anything you want, any being you imagined can be you. You have only to pay the biologic price" (1979, 56). For Di Filippo, however, this price turns out to be a monetary one, just like the price of every-thing else. In the world of *Ribofunk,* prepubescent teenagers can go to the mall to get "any possible alteration on [their] somatype or genotype" (201); adults use enslaved trans-genic hybrids as personal servants (17–33), sidekicks (77–94), gofers (172–87), and even sex partners (20ff); and street gangs create new mind-altering drugs "with their home amino-linkers and chromo-cookers" (87). As long as you've got the cash—or more precisely, the credit—you can have it done. These biological enhancements have nothing to do with utility. They are generally used either as fashion ac-cessories or else as terrorist weapons. Their development does not follow the placid logic of adaptive optimization, but the far crazier rhythms of evolutionary arms races and of commodities proliferating wildly in the marketplace.

The Culture of Real Virtuality. We have seen, in various ways, how the phenomena of virtual reality—from cyberspatial disembodiment to prosthetic hyperembod-iment, and from distributed, artificial intelligence to con-centrated corporeal enhancement—all depend on the fran-tic exchange and circulation of money, or on what Marx (1993b) called the accelerated turnover of capital. Money at once grounds and volatilizes virtual reality. On the most obvious level, money grounds virtual reality, because money is what makes it all happen. A huge influx of venture capi-tal fueled the dot-com bubble of the late 1990s. And when the bubble collapsed, as such speculative manias are wont to do, many of the Internet sites involved suddenly disap-peared. But on a deeper level, the development of virtual technologies coincides with the increasing virtualization of money itself. This is part of the massive reorganization of capital that took place in the last three decades of the twen-tieth century. This transformation has been analyzed in depth by David Harvey, Manuel Castells, and others. Harvey sees the process as a fundamental shift from the Fordism of the mid-twentieth century to a new regime of *flexible accumulation* (147), characterized by just-in-time production, a more frequent use of part-time and temporary labor, and above all "the complete reorganization of the global financial system" due to "a rapid proliferation and decentralization of financial activities and flows" (160–61). There has been "a general speed-up in the turnover time of capital," leading to the increasing "volatility and ephemerality of fashions, products, production techniques, labour processes, ideas and ideologies, values and established practices" (285). Castells adds to this account an emphasis on global networking among corporations of different sizes, as well as among the members of political and financial elites (2000b, 163–215). And he insists upon the novelty and importance, not of information per se, but of technologies that actually use information as their "raw material" (70). Castells also focuses on the synergy between the late-twentieth-century information technology revolution and the growth of globally interdependent finance markets (102ff). These markets are dependent upon advanced information technology, and in return they provide the major impetus for the technology's continuing development. Financial speculation and the electronic media are thus locked together in a mutually reinforcing, positive feedback loop. The more the world becomes a single "global village" (McLuhan and Fiore 1968), the more financial activity mutates into bizarre and extravagant forms (like derivatives[57] and hedge funds), and the more frenetically financial speculation sweeps across the globe, destabilizing entire economies in an instant. Production is subordinated to circulation, instead of the reverse. Money, the universal equivalent, has become increasingly virtual (unmoored by any referent) over the past half century, and everything else is decentered and virtualized in its wake. It's like Derrida on steroids. This delirium is the motor of what Castells calls "the culture of real virtuality": the technologies, communications media, and network structures that provide the material grounding for our increasingly virtual existences (355–406). The physical and virtual worlds should not be opposed; rather, they are two coordinated realms, mutually dependent products of a vast web of social, political, economic, and technological changes.

Out of Time and into Space. The material form of the culture of real virtuality, says Castells, is a new articulation of our experience of space and time. A "space of flows" displaces the familiar "space of places" (2000b, 407–59), while a "timeless time" annihilates traditional cyclical time and industrial clock time alike (460–99). But these transformations are not of equal status. Castells, like many other theorists, argues that "space organizes time in the network society" (407), rather than the reverse. Fredric Jameson also says that postmodern society privileges space over time, in contrast to the ways that modernism, and industrial society, privileged temporality (154). McLuhan, too, contrasts the simultaneity and instantaneity of the new electronic media ("ours is a brand-new world of allatonceness" [McLuhan and Fiore 1967, 63]) to the linear, sequential temporality of print media and the technology of mass production ("duration begins with the division of time, and especially with those subdivisions by which mechanical clocks impose uniform succession on the time sense" [McLuhan 1994, 146]). And Lev Manovich analyzes how the spatial logic of the database has replaced the temporal logic of novelistic and cinematic narrative (218–43). The last word here can be given to Burroughs, who exhorts us to move "out of time and into space"; he likens this transformation to the emergence of animal life out of the oceans and onto dry land.[58] Timeless time—characterized by the negation of duration, the replacement of regular working hours by flex-time and just-in-time production (Castells 2000b, 467), "the breaking down of the rhythms... associated with the notion of a life-cycle" (476), and the overall denial of mortality (481–84)—is not an originary formation, but only the secondary consequence of a deeper mutation: the overall subordination of time to space.

Space Is the Place. The space of flows is virtual space, or cyberspace, or what Jameson, in his discussion of the bewildering architecture of the Westin Bonaventure Hotel in Los Angeles, calls "hyperspace" (38–45). This space can be exhilarating, disorienting, or oppressive, but in any case it is quite different from the space of places. The space of flows has been freed from the constraints of duration: in consequence, distance is abolished, and communication between any two points is instantaneous. Proximity is no longer determined by geographical location and by face-to-face meetings, but rather by global flows of money and information. The predominant form of human interaction in this space is *networking*. I mean this, first of all, in the narrow sense of institutional and professional connections among members of any elite or self-defined group. Samuel R. Delany (1999) lists some of the locations in which net-working traditionally takes place: "gyms, parties, twelve-step programs, conferences, reading groups, singing groups, social gatherings, workshops, tourist groups, and classes" (129). The development of electronic communications tech-nology frees this sort of networking from its dependency upon specific locales and specific times; now networking can flourish on a global scale. In a broader sense, then, net-working just means *being connected:* having an Internet ac-count, or for that matter, a car. For it isn't just online that we encounter the space of flows; we also find it in the fa-miliar American landscape of shopping centers and strip malls, islands of quasi-public (though actually privatized) space in a sea of asphalt, each with its own Starbucks. The logic of the space of flows also organizes the emerging "mega-cities" of the twenty-first century: places like Guangzhou, São Paulo, and Mexico City, where the economy of the First World overlays that of the Third. These places, Castells says, are "globally connected and locally disconnected, physically and socially...Mega-cities are discontinuous constellations of spatial fragments, functional pieces, and social segments" (2000b, 436). This is the situation in which, for instance, gated communities wall out the shantytowns that exist right next door. In contrast to all this stratification, the space of places is what we still think of as homogeneous, continuous physical space. Here, people are brought together by direct physical contiguity, by solidarities (and oppositions) of class, race, ethnicity, and sexual orientation, and by common experiences in their workplaces and neighborhoods. The space of places is often looked back upon nostalgically as a site of "community," where everyone helped their neighbors, and nobody was reduced to the fate of "bowling alone" (Putnam). This is rather a dubious idealization: old-time, small-town communities were just as often characterized by bitter enmities, malicious gossip, and the heavy-handed imposition of censorious, bigoted "community standards." What's crucial about the space of places is rather something other than "community": the fact that, in large urban agglomerations, networking is less important than what Delany, elaborating on an idea from Jane Jacobs, calls *contact*: the serendipitous encounters between strangers—often across class, race, and gender lines—that can change everything (1999, 111–99, and especially 125ff). These sorts of encounters happen most easily in the pedestrian-friendly spaces of older large cities. Delany's main example is New York City's Times Square, before it was "renovated" and gentrified. The space of places is less that of nostalgically idealized traditional communities than that of turbulent urban modernity. It is the space of the urban *flâneur,* memorably evoked by Walter Benjamin (1999, 416–55). And one of the big questions today, facing artists, thinkers, and activists alike, is how to find a twenty-first-century equivalent, within the space of flows, for Benjaminian *flâneurie* and Delanyesque contact.

The Trinity Formula. Castells specifies "at least three layers of material supports that, together, constitute the space of flows" (2000b, 442). Each has its own role to play in the network. The first layer "is actually constituted by a circuit of electronic exchanges" (442). This is the physical basis of virtual communication: the hardware and software, the technology that has to be deployed in order for the network to function at all. This layer is what McLuhan calls the *medium.* If "the medium is the message," as McLuhan famously affirms, this is because "it is the medium that shapes and controls the scale and form of human association and action" (1994, 9). What we say matters less than the fact that we are saying it in the network. This first material layer imposes its shape upon all human expression, but it truly becomes autonomous with the emergence of artificial intelligence. That's when the network itself starts speaking back to us. The second material layer of the space of flows is the geography of its "nodes and hubs...the network links up specific places, with well-defined social, cultural, physical, and functional characteristics" (Castells 2000b, 443). Here the space of flows selectively envelops the space of places, and the virtual digs its hooks into the real. The network itself may be placeless, but access to it is still dependent upon physical location. Not every place is equal. There's a whole series of connections and disconnections. Places can be plugged into the network or unplugged from it. That is how mega-cities get strewn across the landscape, differentiated into heterogeneous zones. The third material layer of the space of flows is the rarefied, homogeneous space in which the global financial and political elite lives, works, and travels. This space is exemplified, for instance, by luxury business hotels, which look pretty much the same all over the world. There's nothing to tell you whether you are in New York City, or Tokyo, or Rio de Janeiro.[59] These hotels all have the same services and amenities, the same monumental, reflecting outsides, and the same sleek, minimalist decor in the rooms, with everything in white or discreet pastels.[60] Overall, this third layer of the space of flows is composed, Castells says, of "residential and leisure-oriented spaces which, along with the location of headquarters and their ancillary services, tend to cluster dominant functions in carefully segregated spaces, with easy access to cosmopolitan complexes of arts, culture, and entertainment" (446–47).

Eden-Olympia. J. G. Ballard's novel *Super-Cannes* is set in and around Eden-Olympia, a high-tech business park on the French Riviera. Eden-Olympia integrates luxurious private residences, efficiently designed corporate offices, up-to-date telecommunications, multilevel parking lots, "artificial lakes and forests" (9), and facilities like gyms, spas, medical services, chic boutiques, and fancy restaurants, all in a single, self-enclosed space. Eden-Olympia is home to "the most highly paid professional caste in Europe, a new elite of administrators, énarques and scientific entrepreneurs" (5). In short, it is "a suburb of paradise" (20)— hence Eden—whose inhabitants can enjoy a godlike detachment from the lesser world outside its gates—hence Olympia. It is also an experiment in social engineering for the new millennium: "the first intelligent city, the ideas laboratory for the future" (115). In all these ways, Eden-Olympia epitomizes the third layer of Castells's space of flows. Everything here is organized in accordance with the requirements of the corporations that run it: "A giant multinational like Fuji or General Motors sets its own morality. The company defines the rules that govern how you treat your spouse, where you educate your children, the sensible limits to stock market investment" (95). As a result, "there's no need for personal morality. Thousands of people live and work here without making a single decision about right and wrong. The moral order is engineered into their lives along with the speed limits and the security systems" (255). And indeed, there is nothing in Eden-Olympia like civil society, or a public sphere, or a democratic government: "an invisible infrastructure took the place of traditional civic virtues . . . Civility and polity were designed into Eden-Olympia . . . Representative democracy had been replaced by the surveillance camera and the private police force" (38). So total a privatization of what used to be the social realm would be oppressive to most people. But it empowers the inhabitants of Eden-Olympia. According to Castells, "the elites form their own society, and constitute symbolically secluded communities." This is because "the more a society is democratic in its institutions, the more the elites have to become clearly distinct from the populace, so avoiding the excessive penetration of political representatives into the inner world of strategic decision-making" (2000b, 446). The reason that so much fuss has been made recently about eliminating government regulation, deciding everything instead by commercial contracts, by the market, and by corporate policy decisions, is precisely because such privatization "basically undermines democracy," as one free-market enthusiast puts it; "it removes behaviors and transactions from the purview of the mob" (May 74). By eliminating democracy, and privatizing both their business relationships and their personal lives, the corporate executives of Eden-Olympia set themselves apart from the general populace. They do not need to trouble themselves with the struggle to survive, or the quest for material wealth, or the desire for self-expression, or any of those other things that everyone else worries over and fights about. From the lofty heights of Eden-Olympia, such concerns are nothing more than petty technical details, easily resolved by corporate engineering.

The Power Elite. Castells says that our corporate rulers are "dominant, managerial elites (rather than classes)" (2000b, 445). What he seems to mean by this is that their power is not founded *directly* upon ownership of the means of production. Rather, it is mediated by their having privileged access to global networks of financial transactions and flexible, lateral connections. The new elites possess the "cultural codes" (446) that permit them (and nobody else) to "ride the light"[61] and to follow the delirious flows of money around the globe. As Ballard puts it in *Super-Cannes,* "the conformist Organization Man of the 1950s is long gone" (95).[62] Instead, "today's professional men and women are self-motivated. The corporate pyramid is a virtual hierarchy that endlessly reassembles itself around them" (96). The inhabitants of Eden-Olympia are continually networking and reinventing themselves, just as management gurus like Tom Peters recommend. Thanks to their mastery of the space of flows and their self-imposed "segregation" from the rest of humanity (Castells 2000b, 447), they have achieved an enviable, posthuman "freedom from morality" (Ballard 2001, 95). Nietzsche's *Übermensch* turns out, Ballard says, to be someone like "Bill Gates or Akio Morita" (297). The twenty-first-century ruling elite is "a new race of deracinated people, internal exiles without human ties but with enormous power" (256). But there is a price to pay for this freedom. The new ruling elites are always busy: "Work dominates life in Eden-Olympia, and drives out everything

else . . . Work is the new leisure. Talented and ambitious people work harder than they have ever done, and for longer
hours. They find their only fulfillment through work" (254).

Rest and Recreation. Such dedication to hard work is
not a problem in itself. But it's difficult to sustain over the
long term. The executives of Eden-Olympia tend to feel
run down after a while. They need some sort of release to
recharge their batteries. And not just any distraction will
do. These executives are too jaded to get a thrill from the
"rather old-fashioned" bourgeois transgressions of "ennui,
adultery, and cocaine" (96). They indulge in these vices,
but more out of habit than for any other reason. Their
pumped-up nervous systems demand something stronger
and fresher. And so, as a kind of therapy, they engage in
ratissages: vigilante street sweeps.[63] Gangs of executives de-
scend at night from the heights of Eden-Olympia to the
surrounding towns and villages, where they systematically
beat up North Africans and members of other racial mi-
norities. They also indulge themselves with "incidents of
deliberate road rage, thefts from immigrant markets, tangles
with the Russian mafiosi" (260). And they recruit homeless
children into pornography and prostitution rings. The point
of this "therapeutic psychopathy" (279) is not to release
deeply repressed violent and sexual impulses, but exactly
the reverse: to implant those impulses into minds that previously lacked them. As Deleuze and Guattari like to say,
the unconscious is not given in advance; it is something
that needs to be actively produced (1987, 149–66). Rather
than being an outpouring of the id, the *ratissages* are deliberately staged exercises: sort of a New Age program of self-
improvement, updating the old Protestant work ethic. And
the therapy works. Once the *ratissages* started at Eden-
Olympia, we are told, the therapeutic "benefits were as-
tounding": the executives felt healthy, and worked harder

than ever, and "corporate profits and equity levels began to climb again" (260). In the new network economy, such controlled outbursts of ritual violence are the motor of personal growth and corporate innovation alike. By going on racist rampages, the inhabitants of Eden-Olympia are able to tap new sources of creativity; they re-establish contact with the outside world from which they have otherwise totally separated themselves.

Big Pimpin'. How does a corporation maintain its edge? How does the informational elite keep up with "business @ the speed of thought" (Gates)? The "therapeutic" *ratissages* at Eden-Olympia are one solution, at least for entrepreneurs and upper-level management. But it isn't easy for a whole organization to stay in touch. As Tom Peters is always warning his readers, there is no progress without failure (54), "standing still is the kiss of death" (102), and professional networking is a "never-ending job" (109). The pace is relentless, and there is no security. Just as Heidegger urges us to live every moment in the awareness of impending death,[64] so Peters suggests that "the best approach to survival is to assume you're about to be (or just have been) laid off, permanently" (97). He tells corporate executives to think in terms of "abandonment" (3), "self-destruction" (27), and "perpetual revolution" (269). A similar dynamic animates the corporate culture of DynaZauber, in Jeter's *Noir*. Harrisch's management style is based on "Harry Denkmann's magnum opus, *Connect 'Em Till They Bleed: Pimp-Style Management™ for a New Century*" (46). The point of Denkmann's system is to control your employees, as pimps do their hoes, by keeping them continually insecure. This is done by making them absolutely dependent on the corporation. It's all "an extension of the old New Orleans whore-hustling motto, that they weren't completely under your control if they still thought they had names of their own" (47). On one hand, the employees must have no resources and no means of support outside of what the corporate network itself provides. On the other hand, the company must distribute its favors so capriciously, arbitrarily, and unpredictably, that the employees can never take them for granted. The name of the game, therefore, is "what the human-resource managers and company psychs called *optimized transience disorientation*" (46). If you can get your workers (up to and including middle-level management) constantly grasping at straws and unsure as to where they will be from one week to the next, then you can guarantee their obedience and extract the maximum possible labor from them. Flextime work (Castells 2000b, 281–96) is often touted as a way to liberate workers from the tyranny of nine-to-five jobs. But of course its actual effect is to increase workers' dependency and insecurity. The same can be said for Peters's boast that "with a bit of imagination (okay, more than a bit), the average job—actually, every job—can become an entrepreneurial challenge" (72). What this means in practice is that workers should be required to approach their customers with the same bright smiles and obsequious enthusiasm that a street pimp expects his hookers to show their johns. When Tom Peters champions things like "challenge," and "reinvention" (252f), and "empowerment" (69ff), he is actually providing the weapons for what Jeter calls the corporations' "psychological warfare" against their own employees (46).

Go with the Flow. Castells sardonically remarks that the ruling elites of the network society live and work comfortably in the space of flows, but they "do not want [to] and cannot become flows themselves" (2000b, 446). No transience disorientation for them; that is the fate that they reserve for the rest of us. The elites know that if they were actually to be swept away in the space of flows, they would no longer be able to rule. The one irredeemable error that a drug lord can make is to become an addict himself. Giving way or letting go—becoming a flow—is a dangerous pleasure, one that the powerful dare not indulge in. They can only let themselves experience it in carefully measured doses, as with the *ratissages* of *Super-Cannes*. As Ballard says, "the rich know how to cope with the psychopathic" (96); they know how to turn such behavior to their advantage, instead of letting it overwhelm them. Nietzsche exhibits a similar sense of caution; barely a page after he celebrates life's "*sacrifice* of its highest types" in "the eternal joy of becoming—that joy which also encompasses *joy in destruction*," we find him urging his "brothers" to "become hard!" in order "to write upon the will of millennia as upon metal" (1968a, 110–12). When Tom Peters exhorts would-be executives to go with the flow, take risks, and act crazy, he disingenuously fails to remind them that hardness and cynicism are equally necessary to success. The actual ruling elites would never make such an error. For they have mastered the true art of the pimp. As Iceberg Slim explains it: "*pretend indifference to enhance desire*... You must learn *strength! Willpower!*"[65] If you want to control your hoes and johns, or your employees and customers, you must always choose profit over pleasure. Calculation must never give way to passion. Every display of emotion is a sign of weakness. Pimps and CEOs, together with free-market economists, may well be the only human beings who actually live according to the passionless dictates of rational choice theory.

Overload. Of course, "optimized transience disorientation" is not only something that happens in the workplace; it is close to being a universal condition in the age of information networks. In Paul Di Filippo's novel *Ciphers*, we are introduced to semiotic (as opposed to biological) AIDS:

"Ambient Information Distress Syndrome. Otherwise known as channel-overload...The victim basically feels overwhelmed by all the information pouring in on him." Symptoms include fugue states and catatonia as the victim "tries to escape all stimuli" (142). But these efforts to retreat are futile. The information keeps on pouring in. In the terminal stage of semiotic AIDS, the sufferer regresses to the form of a fetus and finally blinks out of existence: "his synaptic interconnections became incompatible with the physical structure of our universe, and the cosmos squeezed him out" (268). Similarly, in *Transmetropolitan*, Spider Jerusalem is diagnosed with "I-Pollen Related Cognition Damage": an allergic reaction triggered by exposure to "information pollen." At first, the IPRCD victim suffers from high blood pressure, and occasional nosebleeds, hallucinations, and blackouts. Subsequently, in 98 percent of all cases, there is "continual cognition damage. Memory loss. Intensified hallucination. Eventual motor control damage" (Ellis and Robertson 2003, 36). Semiotic AIDS and IPRCD alike are consequences of Landauer's Principle: the strange fact that, while no energy is needed to create and transmit information, some energy is needed to destroy it. In theory, computation and communication can be perfected to any desired degree of frictionless efficiency. But you cannot get rid of information without dissipating energy in the form of friction or heat. "In handling information, the only unavoidable loss of energy is in erasing it" (Siegfried 74). Memorization is free and easy, but it is entropically costly to forget. Thanks to this basic asymmetry, information cannot be expunged as easily as it is accumulated. At some point, the physical limits, either of the brain's storage capacity or of its ability to dissipate heat will be reached. Think of this as the price of being connected. You can't get rid of old information fast enough, or efficiently enough, to accommodate the new. Your nervous system is pushed to the brink. Eventually, it suffers a catastrophic collapse. It crackles and burns, swamped by information overload, and is swept away in the space of flows.

Out of the Past. McNihil ("son of nothing"), the protagonist of *Noir,* so hates the network society in which he is forced to live that he seeks to expunge it from his vision. Literally: he has had "thin-film insertion surgery" (52), equipping his eyes with an extra optical layer that translates everything he sees into the visual codes of classic film noir. That is to say, McNihil paradoxically uses high technology against itself. At great expense, and thanks to the latest advances in biomechanical interfaces, he is able to obliterate the sight of the "cheap-'n'-nastiverse" (302) that he inhabits. He replaces the surrounding world of corporate logos and weightless simulacra with something equally fictive, a "net of bits and pieces from ancient thriller movies" (302). McNihil sees the world entirely in high-contrast black and white. It always seems to be nighttime. Twenty-first century architecture and clothing styles revert to their 1940s equivalents. But this should not be confused with the "total-environment simulations" of virtual reality (52). McNihil has not retreated into any sort of hallucinatory, imaginary world. He is not a solipsist or an idealist. "Only idiots want to inhabit a world separate from anyone else," he says (55). Rather, McNihil remains a good Kantian. The forms of his experience are indeed constructed by his own (surgically altered) mind, but the contents of that experience, the stimuli to which he responds, come to him from outside, from the common universe he shares with everyone else. There really is a woman sitting on his bed, even if it is only from his peculiar angle of vision that she looks like "a young Ida Lupino," wearing "a period early-forties outfit from Raoul Walsh's *High Sierra*" (42). McNihil's altered vision is thus a case of "seeing the same things that everybody else does, but just seeing them differently" (55). And in the last analysis, "that's the way it is for *everyone*" (55). We all push back against the pressure of the real, trying to impose our own perspectives upon it. This is the process that Nietzsche calls *will-to-power*. With his retinal implants, McNihil simply takes the logic of will-to-power to its extreme. This is not an effort to flee the world so much as it is a way of acting upon it—and being acted upon in turn. For quickly enough, we learn how limited our power actually is. There's the "soft," weightless existence of things that you can "pick up and move around, change and destroy," but there's also the "hard" inertia of "things that really existed, that didn't go away even if you'd wanted them to" (53–54). In the end, "this world is what you can't escape from. It always comes seeping back into your little private existence" (289). No virtual technology can keep it at bay forever. McNihil knows that he cannot actually get away from the network society. His altered vision is rather a kind of shortcut: a means of coping with the horrors of the present by mapping them back onto the fatality of the past. Such a strategy is even evident in Jeter's prose: much of it is written, not just in the past, but in the pluperfect tense, heightening the sense that things are irreparable, that everything is already over. Indeed, the archaic world that McNihil sees is bleak, unforgiving, and filled with menace. Jeter reminds us that noir is a genre grounded in anxiety (241): "the *essence*, the soul of noir" is always "betrayal" (242). The world of film noir was "not a particularly nice world, but [it was] one that McNihil was comfortable living in" (241). For McNihil finds comfort in the very assurance of disaster: when you are trapped, "*at least you know where you stand*. It might be rock bottom, but it was certain. In this world ... there was a certain comfort in that knowledge" (274). Everything else may be subject to doubt, but at least McNihil can count on—and enjoy a bitter satisfaction from—the certainty that, sooner or later, he will be betrayed.

Dark, Annihilating Beauty. What is it that we love about film noir? Why does "this darkly perfect world," as Jeter calls it (302), appeal so powerfully to us? It's not just Jeter and McNihil. Ever since Ridley Scott's *Blade Runner* and William Gibson's *Neuromancer,* the visual and verbal codes of noir have been ubiquitous in dystopian science fiction. It's almost as if we couldn't envision the future without them. The hellish cityscape of Jeter's *Noir* is a direct descendant of *Blade Runner*'s dirty, decaying Los Angeles.[66] Scott does with his film pretty much what Jeter does with his prose, and McNihil with his eyes: he reverse-transcribes the future into the look (oblique lighting, shadows, chiaroscuro, off-kilter camera angles) and feel (urban paranoia, fatalism, exoticism, the femme fatale) of 1940s B movies. A world in which the real has been entirely penetrated by transnational corporations and their technologies of simulation is figured in the form of a dark, grimy, rainy, neon-lit, and overcrowded nighttime cityscape. The darkness is a backdrop, against which the excessively perfect forms of simulation (the icy blonde beauty of the replicants or the alluring smiles of the women's faces on the enormous video billboards that loom over the city) can be most effectively projected. The result is a doubly distanced nostalgia for a lost real: the weightless replication of something that was already a fiction the first time around. And this may have something to do with why betrayal, the inevitable climax of every noir narrative, is comforting to us. Bill Beard suggests that the allure of today's retro noir stylization is that it makes even the most intolerable situations bearable precisely by aestheticizing them, by making them beautiful.[67] Indeed, as Rilke tells us, beauty is nothing but the begin-

ning of terror, which we are only just able to endure; we re-
vere it because it calmly disdains to destroy us. And this
unfeeling transcendence is what seduces us into believing,
along with another poet, that beauty is truth, truth beauty.
Jeter himself nearly says as much. Betrayal is the essence of
noir, and noir, in turn, is the essence of the world, "realer
than real, truer than the dull world beneath the perceptual
overlay" (302). Perhaps, Jeter speculates, McNihil's "surgery had merely been an extraction, the removal of some
kind of invisible cataracts that had prevented [the world]
from being seen in all its dark, annihilating beauty" (241).

Time out of Mind. Philip K. Dick asserts that "real time
ceased in 70 C.E. with the fall of the temple at Jerusalem. It
began again in 1974 C.E.," when higher forces intervened,
for the first time since Jesus' incarnation, to oust Richard
Nixon from office. For all the intervening period, "the Empire never ended"; secular history, culminating in Water-
gate, was nothing but a series of disguised repetitions—or
rather, continuations—of the Roman Empire's persecution
of the primitive Christians (1981, 217). Jeter proposes a similar theory of history. He suggests that "real time had ended
somewhere in the early 1940s" (302), the moment of the
great early noirs. Everything since then has just been a
"shoddy" imitation, like "a curtain made of some flimsy
synthetic fabric," concealing, cheapening, and yet repeating the eternal truths of the genre (302). Today, reality is
just an echo of fiction: "people don't have to" read noir
novels any more, Jeter says, because "now they live in it all
the time" (279). All this is consonant with Fredric Jameson's
sense of the postmodern era as one in which historical time
"remains forever out of reach," evoked only through the
nostalgia of "pop images and simulacra" (25), and in which
there has been "a prodigious expansion of culture throughout the social realm, to the point at which everything in our

social life—from economic value and state power to practices and to the very structure of the psyche itself—can be
said to have become 'cultural' in some original and yet un-
theorized sense" (48). But why are film noir and the 1940s
the privileged sites of this nostalgia? Betrayal may well be
the essence of noir, but this presupposes that there exists,
in the first place, somebody who (or something that) is actually *worth* betraying. The myth of noir is centered on the
figure of the loner, the outsider, or, as Jeter says, the "free-
lancer" (47): someone who isn't part of the system, someone who isn't connected. The male protagonist of the noir
narrative may be noble, or he may be sleazy and corrupt;
in either case, he has his own private code of values, and
he is stubbornly independent of both the cops and the
crooks. He retains this freedom deep within his own mind,
even as external circumstances force his hand and deliver
him to his fate. The noir protagonist's existential isolation
is a virtue, but it is also what inevitably sets him up for betrayal, usually at the hands of a femme fatale. The myth of
noir, then, is really the myth of an autonomous (masculine)
self. And the experience of being betrayed is the test, and
the proof, of that self's authenticity. In this secular world, it
is no longer possible for Jesus to be betrayed and sacrificed
in my place; I have to go through it all myself. The *cogito* of
noir might well read: I am "totally screwed" (272), therefore
I exist.

A Desperate Fantasy. Scott, Gibson, and Jeter all seem
to be reviving the myth of noir in order to use it as a bul-
wark against postmodern emptiness and inauthenticity. In
this reading, science fictional neo-noir would be a last-ditch
attempt to reimagine the space of flows, the pressures of
transnational corporate power, and the ubiquitous technologies of simulation, in more manageable—which is to
say individualistic and traditionally gendered—terms. The

trouble with such a strategy is that, in fact, noir is less an alternative to postmodern simulation than it is itself already an example of it. Film noir first emerged in the early 1940s, but these years were also the time when postmodernity itself was being invented, as Thomas Pynchon makes clear in *Gravity's Rainbow*. In Pynchon's account, World War II was the testing ground for the new electronic media that were subsequently to become the basis of the society of control. The quintessentially postmodern technologies of computers, television and video, and electronically based surveillance and tracking devices, were all initially developed for military purposes, in tandem with that emblematically modernist technology, the rocket. It was on the battlefields of World War II—or more precisely behind the scenes of the war effort—that network power was first articulated and deployed on a massive scale. Beneath the obvious, visible binary opposition of democracy versus fascism, the War was also conducted through a whole range of transversal connections and subterranean collaborations. Large corporations, for instance, burst the bounds of the nation-state, as they sold weapons to both sides and found opportunities, thanks to all the destruction and chaos, to pursue their own agendas of product development and eventual market saturation. It was in the crucible of the War that these corporations first learned how to become transnational, in the sense that has become ubiquitous today. More generally, the pages of *Gravity's Rainbow* are filled with paranoid apprehensions, crazed behaviorist schemes, strange correlations at a distance, phase shifts, feedback resonances, and other nonlinear transformations that defy prediction. Pynchon's point is that all these "network effects" were indeed produced by and through World War II, even if they were invisible to observers at the time. Back then, such operations could only be registered indirectly, in the form of widespread, unanchored feelings of dread, fatalism, melancholy, and hopeless romanticism. This is, of course, precisely the emotional atmosphere of film noir. *Gravity's Rainbow* suggests, therefore, that the noir sensibility is both a symptomatic expression of the nascent network society, *and* a heartfelt, although futile, attempt to negate it. Even back in the early 1940s, the noir myth of the doomed existential self was little more than a desperate, and secretly wishful, fantasy. If noir today is an ironic anachronism, this is because it already was one the first time around.

The Pseudoromantic Mystique of Film Noir. Film noir generally seems to unfold in the pluperfect tense. That is to say, it refers to a time that has always already been revoked, a past that was never actually present. Film noir skips over the present moment entirely, conflating the time of fatality (the blocking, or the closing-down, of the future) with the time of nostalgia (the fixation upon an ever receding past). This built-in obsolescence may be the reason why we are so in love with film noir today, why we endlessly revive and recycle it. In a world of simulacra, a world where "the real is no longer possible" (Baudrillard 2001, 180), noir is the privileged simulacrum of the absence of simulacra; it signifies, and simulates, the lost real itself. It makes us feel (even though we know better) that this real at least *used* to exist. The appeal of a neo-noir film like *Blade Runner* is that it gives a new life (albeit an artificial and time-limited one) to old-fashioned humanist pathos. *Blade Runner* depicts a wholly dehumanized and de-organicized world; at the same time, it humanizes its replicants (artificial humans) by endowing them with a deep awareness of transience and mortality, of "moments ... lost in time like tears in rain." Even though the replicants are simulacra themselves, they nonetheless feel our pain at finding ourselves in an entirely simulated world. This reassures us that, after all, nothing has really changed. The fatality of the future is safely pack-

aged in the nostalgic colors of the past. Mourning the supposed loss of the real (as Baudrillard and *Blade Runner* both
do) is actually a way of preserving it unquestioned. The
melancholy of *Blade Runner* is, in Borges's (332) phrase, one
of those "apparent desolations and secret consolations" with
which we delude ourselves in order to deny the irreversible
passage of time. The more gritty and downbeat noir style
might seem to be, the more it becomes hip, chic, and fashionable. The real lesson of *Blade Runner* is how easy it is to
commodify fatality and nostalgia. It isn't all that much of a
step from Scott's film to the "Noir Center" in Hollywood,
as imagined by Thomas Pynchon in his novel *Vineland* (326).
The book is set in 1984, and the Noir Center is a shopping
mall designed around film noir motifs, with "an upscale
mineral-water boutique called Bubble Indemnity, plus The
Lounge Good Boy patio furniture outlet, The Mall Tease
Flacon, which sold perfume and cosmetics, and a New
York–style deli, The Lady 'n' the Lox." This is "yuppification
run to some pitch so desperate," Pynchon writes, that it
drowns in its "increasingly dumb attempt to cash in on the
pseudoromantic mystique" of old movies.

New Rose Hotel. William Gibson's cyberpunk science
fiction also flirts with the use of noir motifs as fashion ac-
cessories. Much of the allure of Gibson's books comes from
his noir-inflected prose, mixing hard-boiled concision with
an eye for tellingly emblematic details, as in the famous
description that opens *Neuromancer:* "the sky above the
port was the color of television, tuned to a dead channel" (3).
The allure also comes from Gibson's characters, who are
generally too hip for words: tough, desperate, nihilistically
disaffected—and secretly, beneath it all, hopelessly ultra-
romantic. Gibson thus seeks to update the old noir arche-
types, rather than merely recycling them for their retro
appeal as Ridley Scott does. But Gibson also places his

doomed individualists in a world that has little room for them, a world very different from that of classic noir. Gibson takes the full measure of a society driven by biotechnology, advanced computing and communications systems, and warfare among corporations (rather than among nation-states). You can see this even (or especially) in the short story "New Rose Hotel" (Gibson 1987, 103–16), the most explicitly noirish text Gibson has ever published. The world of this story is dominated by "the zaibatsus ... the multinationals," whose lifeblood "is information, not people. The structure is independent of the individual lives that comprise it. Corporation as life form" (107). The zaibatsus are hive entities, superorganisms with their own transhuman lives. The world they dominate is no longer a place of dark conspiracies, paranoid apprehensions, and existential leaps of faith, for the old noir landscape has been replaced by a thoroughly networked one, where even conspiracy, terror, and danger have been routinized and bureaucratized. "New Rose Hotel" is a classical noir tale of love and betrayal—but one that has been mutated to fit this new world. The unnamed narrator and his associate Fox hire Sandii—the femme fatale of the story—to seduce Hiroshi, a top research biologist. The aim is to get Hiroshi to defect from one zaibatsu (Maas) to another (Hosaka). The scheme unfolds smoothly at first, but the narrator makes one big mistake: he falls in love with Sandii himself. In noir fiction, this is always a fatal move, for of course Sandii double-crosses the narrator, with calamitous results for all concerned. She reprograms Hiroshi's DNA synthesizer so that it unleashes a "meningial virus" (116) custom-made by Maas. This virus kills not just Hiroshi, but Hosaka's entire cutting-edge research team. The narrator realizes that he should have known; the last night he spent with Sandii, he had found an unlabeled diskette in her purse (110). But now it's too late. "New Rose Hotel" is narrated retrospectively, as if by a dead man. Hiroshi is dead, Fox is dead, Sandii has disappeared, and the narrator himself is just waiting to be killed. He replays the story over and over again, inside his head. But even though he knows that Sandii has betrayed him, he cannot bring himself to give her up. Gibson's prose is dense with melancholy and yearning, as the narrator willingly embraces the catastrophe his life has become.

The Edge. In "New Rose Hotel," Fox and the narrator are always looking for what they call "the Edge"—with a capital E (103). The Edge is "that essential fraction of sheer human talent, non-transferable, locked in the skulls of the world's hottest research scientists" (103). Fox and the narrator are interested in Hiroshi because he is someone who has the Edge: he's "a freak, the kind who shatters paradigms, inverts a whole field of science, brings on the violent revision of an entire body of knowledge" (108). In the course of the story, the Edge takes on a mythical status. That is because it already *is* a myth of our society, in several senses. In the first place, it is the Idea of Genius. As Kant, who did more than anyone to codify the notion, puts it, "genius is a *talent* for producing something for which no determinate rule can be given" (1987, 175). The Edge is the faculty of working outside of all maps and rules and creating something that is absolutely unforeseeable. It's the tiny margin that makes for innovation, "the last delta-t" (Pynchon 1973, 760), the "difference that makes a difference" (Bateson 459). But, in the second place, the myth of the Edge is one of recognition as well as creation. An inventor who toils in obscurity, or a poet whom nobody reads, doesn't have it. Such a person is not a genius, but a crackpot. In contrast, whoever has the Edge is hip and cool. Not only does he or she not follow fashions, he or she sets the fashions that are then followed by others. The Edge is that charisma that energizes an otherwise entropic and routinized world. In the third place, and following from the second, the myth of the Edge has an economic sense. It is also a basic myth of capitalism. The entrepreneur revitalizes the market, rescuing the economy from inertia. If you are the one who creates the fashions, then you can make money from them. If you have the Edge, you can outdistance your competitors and come up with "basic patents" that will earn you "tax-free millions" (Gibson 1987, 108). There is always some degree of slippage among these three different meanings. Having the Edge, as Maas and Hiroshi have it, means moving from the talent of innovation, to the mystique of charisma, and finally to the accumulation of profit. And admiring the Edge, or searching for it as Fox and the narrator do, conversely means moving from the mere quest for big bucks to a more mystical sort of dazzlement. Ultimately, the narrator is less concerned with making money from the Edge than he is with being somehow redeemed and reinvigorated by its touch. Fox compares it to "a ritual laying on of hands" (114). In all these senses, the Edge is a singularity. It is something that comes from outside the network, something that, miraculously, isn't connected. And this may explain the desperation with which Fox and the narrator search for it. But even though the Edge is somehow apart from the network, it also cannot avoid being finally reabsorbed by it. If Maas cannot make a profit from the Edge, then they will make sure that nobody else can, either. And this is what the narrator fails to understand. Instead of finding the Edge, he only finds love—and with it, betrayal. His failed love is a kind of dark shadow or negative of the Edge. It's the only form that a radical exception can take in a totally networked and commodified world. And that is the meaning of noir, when everybody is connected.

Delectatio Morosa. Abel Ferrara's film adaptation of "New Rose Hotel" carefully follows the plot of Gibson's short story, even down to the tiniest details. But rather than making a thriller or an action picture, Ferrara pushes all the exciting events of the story offscreen. And rather than dwelling on the story's high-tech premises with spectacular special effects, he shows us nothing more advanced or magical than video feeds on the screen of a Palm Pilot. *New Rose Hotel*, like Godard's *Alphaville* a third of a century earlier, finds its science fiction technologies and settings entirely in the corporate architecture of the present day.[68] Many of the film's scenes are set in anonymous, expensive hotel rooms. These rooms are luxurious, but they feel cold and impersonal; they have lots of direct light and empty space, and very little furniture. They have clearly been built for control and efficiency, rather than for pleasure or even comfort. Other scenes take place in murky and cluttered public spaces; these are lit in a more recognizably noirish manner, with neon reds or blues limned against the darkness. There are few establishing shots in the film, and the space is often fragmented or parceled out irregularly, with incongruous closeups and eyelines that do not match. Recording devices and screens are everywhere, and much of the film consists of grainy, reprocessed video footage. Indeed, the important plot events are exclusively conveyed through blurry video fragments and off-camera telephone voices. The protagonists, Fox (Christopher Walken) and X (Willem Dafoe, corresponding to the short story's unnamed narrator), mostly just respond to the information they receive. They both seem to be spinning in a void; Dafoe broods passively in shot after shot, while Walken makes florid pronouncements, gesturing theatrically with his cane. Fox and X would like to believe that they have the Edge and that they are initiating the action. Yet they are helpless in the face of betrayal. They find themselves always behind the curve, reacting to events that they are unable to control, anticipate, or even quite comprehend. This predicament is shared by the audience. In *New Rose Hotel*, the world has lost its solidity and retreated into its images. But if the forces that Fox and X must deal with are vague and impalpable, impossible to pin down and represent, this does not make them any less murderous. Mess with Maas or Hosaka and you're dead meat. Betrayal and death are still the end of every story; it's just that their glamorous allure has faded. In Ferrara's morbid take on film noir, the genre's desperate (and secretly comforting) romanticism is replaced by a jaded, decadent aestheticism.

What Do Women Want? The femme fatale is the key to noir mythology. The noir narrative could scarcely exist without her, for only she can put the male protagonist to the test and thereby validate his existence. In equal parts dangerous and alluring—or dangerous precisely because she is so alluring—the femme fatale is the ultimate prop for male fantasy. Even when she becomes the protagonist's ally, rather than his tempter or betrayer, she is still a mystery and a source of trouble. Her lurid sexuality is a danger, enticing the male protagonist to drop his guard. Even more perturbing is her combination of stereotypical ultrafemininity with the ostensibly "masculine" traits of toughness and ruthlessness. This category confusion is yet another reason why the male protagonist feels isolated and lost. It's not that he really believes in those old gender stereotypes any longer, but he has no other framework for understanding sexuality and desire. Now, however much Gibson's "New Rose Hotel" revises noir mythology in other respects, it remains quite traditional in its treatment of the femme fatale. It takes Abel Ferrara to reimagine the character of Sandii—played in the film by Asia Argento. Ferrara replaces noir's old-fashioned feminine mystique with a sensibility straight out of softcore porn. In one scene, Sandii takes X to a sex show featuring a man and two women; she joins into the action, making out with one of the women, all the while encouraging X to watch. In a later scene, Fox throws a party for X, with four giggling geisha at his disposal. (X lies passively on the hotel bed with all four women writhing over him at once, while Fox videotapes the proceedings.) There are also plenty of cheesecake shots of Argento herself, especially emphasizing her breasts and the elaborate tattoo on her belly. I doubt that I would be able to convince anyone that this is somehow a feminist gesture on Ferrara's part, but at the very least, it is a mutation in the usual misogynist dynamics of noir. When the woman thus presents herself, without a qualm, before the camera, she is no longer playing the old game of enchantment and seduction, of what Baudrillard calls "the secret and the challenge" (2001, 162–68). Rather, she exemplifies that postmodern hypervisibility, or obscene transparency, that Baudrillard finds so horrifying: "the body is already there *without even the faintest glimmer of a possible absence*, in the state of radical disillusion; the state of pure presence" (1988, 32). I think that Baudrillard describes the situation well, but we need not recoil from it as he seems to do. For what this means is simply that the old mystery of "femininity" has collapsed; the female body is no longer a mere prop for male fantasy. It is too overwhelmingly present, too *visible*, for that. This is indeed a "state of radical disillusion"; whatever sleazy pleasures Ferrara (and the male heterosexual viewer) may be getting off on, the myth of the femme fatale is no longer part of it, for in the realm of hypervisibility, the woman is not any more (nor any less) mysterious than the man. If Sandii's intentions and motivations remain inscrutable in Ferrara's film, then to an equal extent, so do X's. The story and the screenplay present Sandii as a woman of mystery, but in the movie Argento's easy compliance without commitment is evenly matched by Dafoe's

passive, fatalistic blankness. Ferrara implodes the traditional gender dynamics of noir, for he understands that, in
the age of what Baudrillard (1988, 33 and 35–36) rather
hysterically calls "the excessiveness and degradation of the
image . . . this promiscuity and the ubiquity of images, this
viral contamination of things by images," such dynamics
are no longer relevant. And a good thing too.

You Can't Put Your Arm around a Memory. In Abel
Ferrara's adaptation of "New Rose Hotel," the story ends
long before the film does. The entire plot unfolds in an hour
and seven minutes. It ends with X taking refuge in a "coffin
hotel" just outside Narita airport, waiting for the hit men
to come and get him. It would seem that the movie is just
about over. But instead, there is a long sequence in which,
literally, nothing happens. For twenty-two minutes, we see
X lying in this small room, just pondering, and poring over
a few objects taken from Sandii's purse. Closeups of Dafoe
are intercut with flashbacks from earlier in the film, together
with some video surveillance footage. The flashbacks are
wide ranging and not presented in linear order. They include both literal repetitions from earlier in the film and
many subtle variations. There are alternate takes of scenes
we have previously seen: the camera is at a different angle,
or the dialogue is slightly different, or the positions of the
actors have been inverted. Also, at times, the soundtrack
from one scene accompanies the visuals from another. In
addition, there is some entirely new footage: action that belongs to, or extends from, scenes we saw earlier in the film,
but that was omitted the first time around. In this way, Ferrara actualizes memory, with all its uncertainties and its
underlying futility. For memory is a kind of virtual reality.
An event is entirely singular; it only happens once. But we
do not remember an event only once; as it reverberates in
our minds, it is continually being inflected and rearranged

according to the vagaries of desire. This is why storytelling is never reliable; and also why, as Jean-François Lyotard famously maintained, in the postmodern world there is no Grand Narrative that explains everything, but only a multitude of competing small narratives. You might say that Ferrara "postmodernizes" Gibson's modernist narrative strategy. Gibson has the narrator telling the story in retrospect; Ferrara dramatizes this retrospection, making it an additional part of the story. Where Gibson separates the act of narration from the events being narrated, Ferrara puts them both on the same plane. Ferrara dissolves Gibson's "split subject" into immanence; the act of reflection is no longer separate from the "promiscuity," the ongoing, chaotic flux of images and sounds being reflected upon. In this way, Ferrara creates a mood of free-floating melancholic disillusion that is quite different from the existential anguish of the classical noir protagonist, as well as from the nihilistic "cyberpunk" attitude that we find in Gibson's updating of noir.

The Dead Man. In Jeter's *Noir*, it always comes back to the bedrock situation of betrayal. Through surgery, McNihil has made his world conform to the look and feel of film noir, but as Turbiner, an old writer of noir fiction, reminds him, "the look, all that darkness and shadow, all those trite rain-slick streets—that was the least of it. That had nothing to do with" the inner truth of noir (242). Since noir is about betrayal, it follows that its appearances will always be deceiving. And indeed, it turns out in the course of the book that Turbiner himself is the one who betrays McNihil—a nasty payback for McNihil's zealous efforts to enforce Turbiner's copyrights. The final truth of noir, then, is that we cannot hang onto anything, not even to the atmosphere of noir itself. McNihil is a noir character because he is (as his name implies) a man entirely without attachments. He is

betrayed by the only person he trusts, just as he has an obscure feeling of guilt for having betrayed the only person
who ever trusted him—his wife, now dead. And what is
worse, he betrayed her for a phantom. Throughout the
novel, McNihil is obsessed with a virtual femme fatale
figure, ironically named Verrity. To McNihil, Verrity is "realer
than real" (147); she, and nobody else, is the "one that de-
cides [his] fate" (124). McNihil has betrayed his wife for
Verrity, in thought if not in deed, and she is the only one
who has ever actually defeated him. And yet, toward the
end of the book, McNihil is finally forced to admit to himself that Verrity "never existed. She was a lie. A fiction."
She doesn't even exist in the Wedge, that kinky virtual realm
where flesh and fantasy intermingle. McNihil himself "made
Verrity up," he now realizes, to explain and justify his own
failures (408). Better to be able to blame your disaster on a
grand passion, or the machinations of a femme fatale, than
on a banal string of everyday lies, errors, and omissions.
Better paranoia, as Thomas Pynchon puts it, than "anti-
paranoia, where nothing is connected to anything, a condition not many of us can bear for long" (1973, 434). With
McNihil's recognition that he invented Verrity, the entire
fantasy structure of noir collapses. It's precisely because
the noir hero's self has so little actual substance that it
needs to be validated, and made concrete, through betrayal. But in the "radical disillusion" of the network society, where everyone connects everyone else, even betrayal
isn't enough. It is too commonplace; everybody does it.
Something stronger, and more extreme, is necessary. And
so Jeter fulfills the noir mythology by making his hero literally into a dead man. For it is only by killing himself, and
then returning as a zombie, that McNihil is able to thwart
the plans of Harrisch and the DynaZauber Corporation
and bring the novel to a suitable genre conclusion.

Indeadted. Why does McNihil come back to existence as a zombie? It's all because he owes a lot of money. In the network society, where every human transaction is instantly monetized and commodified, indebtedness is what makes it all work. You are only as good as your line of credit. "The problem with being in hock," as Jeter says, is that "it [gives] other people power over you, and not the enjoyable variety" (165). But such a condition is unavoidable; you pretty much have to go into debt to function in society at all. Only by borrowing can you build up a credit rating. And your credit rating is the most important factor in defining your rights and duties. In the industrial, disciplinary era, citizenship was defined by the right to vote. But today, voting and citizenship are irrelevant; all that matters is purchasing power. Personhood is predicated upon Visa and/or MasterCard. Instead of voting, consumers "reveal" their "preferences" in the marketplace. As Deleuze says, "marketing is now the instrument of social control... A man is no longer a man confined but a man in debt" (1995, 181). There's no longer any need to lock me up in order to regulate my behavior; I can be controlled much more easily through my financial records—which have the added advantage of tagging along with me, no matter where I go. Of course, this new fluidity does not mean that incarceration is eliminated in the control society; rather, it is increasingly privatized as criminality becomes a function of credit status. Octavia Butler, in her novel *Parable of the Sower*, envisions the institution of *debt slavery*: "people were not permitted to leave an employer to whom they owed money. They were obligated to work off the debt either as quasi-indentured people or as convicts. That is, if they refused to work, they could be arrested, jailed, and in the end, handed over to their employers... Worse, children could be forced to work off the debt of their parents if the parents died, became disabled, or escaped" (295). Jim Munroe depicts a similar situation in his novel *Everyone in Silico:* if you owe money, and your debt is purchased by "Sony Holdings Ltd" or "Microsoft America," then you can be forcibly "relocated to a factory of their choosing" to "work off said debt" (93, 197–98). Jeter follows this process to its logical culmination. In the world of *Noir,* you cannot forget your debts for even a second. The principal and interest figures are always running across your palm, numbers in red "like legible, luminous blood" (167). And there's no way, ever, to cancel this debt, no procedure for declaring bankruptcy and starting anew. Your creditors will even pursue you in death, resuscitating you to work as a zombie to discharge your debt. Such is McNihil's case: after he kills himself, he is brought back to the living world as one of the "indeadted" (167, 474, 475): "walking corpses" (474) doomed to wander the earth, working incessantly, until they have paid back every last penny they owe.

Debt and Death. Debt and death are intimately connected—and not just because the words sound alike in English. Nietzsche speculates that our very idea of justice grew out of "the oldest and most primitive personal relationship, that between buyer and seller, creditor and debtor" (1969, 70). Moral obligation is simply a metaphorical extension of the material obligation of repayment. And punishment arose as a way of enforcing this obligation; it is a way of compelling repayment, usually with interest. "The lawbreaker is a debtor who has not merely failed to make good the advantages and advance payments bestowed upon him but has actually attacked his creditor," and therefore needs to be treated accordingly (71). Capital punishment is the ultimate form of this compulsory debt collection. If I cannot discharge my debts in any other way, then I can satisfy my creditors by forfeiting my life to them. Burroughs

draws a similar connection. He recalls how "the old novel-
ists like Scott were always writing their way out of debt";
analogously, he suggests, the modern novelist "sets out to
write his way out of death" (1987, 3). It's an endeavor that
always fails in the long run; eventually the writer has
"reached the end of words, the end of what can be done
with words. And then?" (258). Death, Burroughs says, "is
equivalent to a declaration of spiritual bankruptcy" (3), a
final wiping-out of all your accounts. For both Nietzsche
and Burroughs, death is the final term in the progression of
debt. The absoluteness of death makes it an equivalent for
that which otherwise has no equivalent: an interminable
series of payments. For the creditor, the debtor's death offers a pleasurable recompense, in lieu of the funds that will
never be collected. And for the debtor, death at least offers
a way out, an escape from the hell of unending obligations.

Life-in-Death. But in the network society, the relationship between debt and death is inverted. Rather than death
resolving and canceling debt, debt extends the moment of
death interminably. The "bad infinity" of debt—and especially of continually accumulating compound interest—replaces the finitude and finality of death. You are no longer
able to put an end to things by committing financial or actual suicide, by declaring material or spiritual bankruptcy.
That loophole has been eliminated. The postmodern condition, delineated in *Noir*, is rather "to be in debt ... to be so
far into the hole, owing so much money, that your own
death wouldn't get you off the hook" (474). Jeter's indeadted are not really alive; they are affectless, without
pleasures or passions, without the distractions of "memory, dreams, hope" (110). They no longer experience that
"pure mercantile hunger" (107) that motivates the living in
an age of informational capitalism. But they are also de-
prived of the peace of the grave. They can never rest. They

do nothing but work, and they subsist only in order to work. What's worse, this work never comes to an end: the earnings of the indeadted are usually not enough even "to service the interest on whatever debt load they had died carrying," let alone to pay down the principal (111, cf. 476). In the world of *Noir*, the indeadted do not actively suffer in the way that copyright violators do, for they are as insensitive to pain as they are to pleasure. But they *do* have to endure the misery of Coleridge's Life-in-Death (Jeter 110), or the subtle metaphysical torment described by Blanchot: "the sickness in which dying does not culminate in death, in which one no longer keeps up hope for death, in which death is no longer to come, but is that which comes no longer" (103). Jeter describes a situation in which the "border phenomenon" separating life from death, or humanity from "insentient matter," has been "erased"; the living dead are just "a point on a continuum that [runs] back down into the trash and rubble filling the streets and burnt-out building husks" (105).

I Can't Go On, I'll Go On. In the network society, you are what you owe, instead of what you eat. Deleuze distinguishes the old regime of confinement from the new regime of debt in this way: "In disciplinary societies you were always starting all over again ... while in control societies you never finish anything.... Control is short-term and rapidly shifting, but at the same time continuous and unbounded, whereas discipline was long-term, infinite, and discontinuous" (1995, 179–81). It is worth going over this contrast in detail. First, debts are "short-term and rapidly shifting," because they are always being rolled over and renewed. When you're in debt, "things can change on you. Really fast," since the turnover of capital is so rapid, and speculation-driven markets are so volatile (Jeter 167). The control society "is based on floating exchange rates," whereas the disciplinary society was grounded in the long-duration fixity of the gold standard (Deleuze 1995, 180). The only thing that doesn't change in the control society is the brute fact that you *are* in debt; you will never finish with that, never be able to clear the ledgers and start all over again. Second, when everything is coded in terms of debt, the old disciplinary separations and discontinuities—between what is human and what is not, for instance—tend to break down. The human is now connected with everything else, across the lines of Jeter's erased border, or Donna Haraway's "leaky distinctions." The great opposition between life and death gives way to a situation in which different degrees of death, like different degrees of debt, are spread all along a "continuum." There are no longer any closed spaces or pre-given categories; instead, fine adjustments are continually being made, through the process of "universal modulation" (Deleuze 1995, 182). Third, the disciplinary society was "infinite," while the control society is "unbounded" (*illimité*). To be confined means to be strictly bounded in space and time. But confinement also has an infinite dimension, since only an absolute and transcendent power (the State as arbiter, the impartial judiciary, or the Kantian moral law) has the inflexible authority to decree it. Debt, in contrast, is immanent, contingent, and flexible; it is contracted in varying circumstances and amounts, it is always renewable, and it is managed through adjustable interest rates (Jeter 167). Debt has no claims to transcendence, but for that very reason, it is potentially limitless, propagating onward and outward indefinitely, without bounds.

Children of Production. Sometimes we think of monsters as archaic beings, oozing out of our primordial imaginings, resistant to the valorizations of capital and the rationalizations of modern science and technology. Other times we see them as uncontrollable by-products of technology run amok, like Frankenstein's monster or Godzilla. But both of these approaches ignore the ways that monsters are intrinsic to the ordinary, everyday reality of capitalism itself. There is nothing extraordinary or supernatural about the indeadted in *Noir*, for instance; indeed, Jeter says, the reanimated dead are "wired into cold reality, in a way that the living could never be" (110). The indeadted are immanent to social reality, rather than invading it from outside. More generally, both vampires and zombies are vital (if that is the right word) to the functioning of capitalist society. Traditional Marxist theory, of course, focuses on vampires. Marx himself famously describes capital as "dead labor which, vampire-like, lives only by sucking living labor, and lives the more, the more labor it sucks" (1992, 342). The vampire grows, not through any productive activity of its own, but by expropriating a surplus generated by the living. But this is only one side of the story. No predator or parasite can survive in the absence of prey. Just as every pattern of "information" needs to instantiate itself in some sort of tangible medium, so vampire-capital can only extract its surplus by organizing its legions of zombie-labor. And as capitalism progresses and expands, these legions must be organized on an ever greater scale. The nineteenth century, with its classic regime of industrial capitalism, was the age of the vampire. But the network society of the late twentieth and twenty-first centuries is rather characterized by a plague of zombies. As Deleuze and Guattari put it, "the only modern myth is the myth of zombies—mortified schizos, good for work, brought back to reason" (1983, 335). And again: "the myth of the zombie, of the living dead, is a work myth and not a war myth" (1987, 425).

The Tendential Fall of the Rate of Profit. The prevalence of zombies in the postmodern world—from George Romero's living dead to K. W. Jeter's indeadted—

is one consequence of the process that Marx calls the "tendential fall of the rate of profit." In the course of capitalist development, Marx says, "the mass of living labour applied continuously declines in relation to the mass of objectified labour that it sets in motion" (1993a, 318). That is to say, businesses expand by continually accumulating capital. More and more living labor is transformed into dead labor, through the extraction and realization of surplus value, and the zombification of the work force. Productivity increases, and prices are driven down, because the same amount of living labor is progressively able to produce more commodities by setting more dead labor into motion. Yet there is a contradiction hidden within this process. "The devaluation of labour-power," as David Harvey says, "has always been the instinctive response of capitalists to falling profits" (192). But this is a short-term solution that, in the longer term, only worsens the problem, for the dead weight of accumulated zombie labor acts as a sort of brake on the continued valorization of capital. What Jeter calls "the entire economy of the dead" (110) implies an ever-dwindling rate of return. As more and more dead labor is amassed, the ratio between living labor and dead labor, and therefore the *rate* at which profit can be realized, declines, tending toward a limit of zero. In *Noir,* as the bodies of the dead decay, their skills become so "low-grade" that they are only able to work as scavengers or "shambling scrap-picker[s]." Garbage themselves, "like crows minus even a bird's intelligence," all they are capable of is sorting through garbage, "rooting around for scraps of aluminum foil, the still-shiny tracings off busted circuit boards" (111). Gathering and recycling this detritus, the indeadted work so feebly and inefficiently that they will never earn enough to pay off their debts. In turn, this means that the corporations employing them will never realize their full potential profit. While it is still cheaper to employ the dead in this way, "than spend the money for automated scanning machinery to do the same thing" (111), the difficulty of realizing profits remains. The result is overproduction and unsold goods, which in turn increases the amount of garbage for the dead to sort through. Now, for Marx the "tendential fall" is precisely that: a tendency, and not an inevitable "law." It is always accompanied by countervailing tendencies. An increase in productivity means a diminished rate of profit, if the rate of exploitation stays the same. But increases in productive efficiency often also involve a concomitant intensification of the rate of exploitation as well. Also, the relative decline in the *rate* of profit is mitigated by an absolute increase in the overall *quantity* of profit, since new productive technologies generally enable a widening of the scope, and an acceleration of the very process, of capital accumulation. As Deleuze and Guattari put it, "capitalism confronts its own limits and simultaneously displaces them, setting them down again further along" (1987, 463). A positive feedback loop is thus set into motion: the accumulation of profit leads to the decline in the rate of profit, which in turn spurs an even greater absolute accumulation, which in turn leads to an even greater relative decline, and so on ad infinitum. At the tendential limit, nearly every last person in the world will have become a zombie. "All you really need," as Harrisch of DynaZauber says, "is enough of an uninfected elite at the executive level to rake off the profits"; everybody else can just flounder in eternal indeadtedness (463).[69]

The Call of Cthulhu. Marx associates the process of vampirism with capitalism, but the actual *figure* of the vampire is better seen as a precapitalist throwback. Vampires are aristocratic and not bourgeois. They do not accumulate, but expend. Their rapacity is sexually charged; it springs from passion, rather than economic calculation. In a close analysis of Bram Stoker's novel, Friedrich Kittler (50–84) shows how Count Dracula is defeated by the science and technology of the late nineteenth century and particularly by the new "mechanical discourse processing" devices of the time, the typewriter and the phonograph (74). Thanks to these devices, the vampiric flow of blood is reduced to "nothing more than a flow of information" (77). As such, it can be calculated and manipulated, just like any other packet of bits. This is what allows the vampire hunters to capture and destroy Dracula; in a world homogenized by the commodity form, and by money and information as universal equivalents, "the Other no longer has a place of refuge" (82). Even in 1897, at the moment of his first appearance in England, Dracula is already obsolete. In Stoker's novel, he falls victim to the Western European colonialist enterprise (82). This also means that the Count is captured by what Marx calls the *formal subsumption* of labor under capital: one of the "processes whereby capital incorporates under its own relations of production laboring practices that originated outside its domain" (Hardt and Negri 255). Dracula is a feudal master and not a capitalist, but for all that, his ravages are captured by, and brought within the scope of, the capitalist world market. Today, things are even sadder for the old Count. More than a century after his initial encounter with the technologies of capital, Dracula only survives as a retro fashion icon. This is because we have now reached the point, as Michael Hardt and Antonio Negri suggest, when "the subsumption is no longer *formal* but *real*. Capital no longer looks outside but rather inside its domain, and its expansion is thus intensive rather than extensive . . . Through the processes of modern technological transformation, all of nature has become capital, or at least has become subject to capital" (272). Vampires today are popular commodities, probably more popular than ever, but they are not really terrifying any longer. Rather, they feed our naive hunger for a safe dose of exoticism. They are commodities that ironically embody our nostalgic yearning for a time when not everything had yet been subsumed under the commodity form. In the network society, the monstrosity of capital cannot take the overly cozy and comforting shape of the vampire. It must be figured as something absolutely inhuman and unrecognizable: Lovecraft's Cthulhu, rather than Stoker's Dracula.

Slake-Moths. China Miéville's novel *Perdido Street Station* turns on the terrifying figures of the *slake-moths:* predatory psychic vampires who feed on human thoughts and dreams. These creatures are Miéville's answer to Cthulhu. Physically, the slake-moths are more-than-human-sized carnivorous insects: they metamorphose from brightly colored grubs into creatures with vast wingspans, three pairs of arms or legs with claws, and "a huge, prehensile slavering tongue" (362). But their rapacity is mental, not physical. They hypnotize their victims with psychedelic patterns of light and color running across their wings. And then they slip their tongues into the heads of their prey, and drink, "not the meat-calories slopping about in the brainpan, but the fine wine of sapience and sentience itself, the subconscious." The slake-moths gorge themselves on "the peculiar brew that results from self-reflexive thought, when the instincts and needs and desires and intuitions are folded in on themselves, and we reflect on our thoughts and then reflect on the reflection, endlessly..."(375). The victims' minds are sucked dry; their bodies are left behind in a zombie state, still breathing and metabolizing, but without awareness or voluntary motion. In this way, Miéville reconfigures the entire economy of vampires and zombies (though he rarely uses either word). Zombies are no longer alienated workers, producing value but excluded from its enjoyment. Instead, they are already-exhausted sources of value, former vessels of creative activity and self-reflexivity that have been entirely consumed and cast aside. No longer capable of living labor, they are not a renewable resource. It is true that in a certain sense the slake-moths and their prey form a "a little ecosystem . . . a perfect loop" (375). For after the moths drink their victims' dreams, their feces fertilize and feed our nightmares, which in turn makes us all the more delicious as slake-moth prey. It's just "like . . . rabbit-shit that feeds the plants that feed the rabbits"(375). But such an ever amplifying, positive feedback cycle does not settle down into any sort of equilibrium. As long as the slake-moths are alone "at the top of their food chain . . . without predators or competitors" (379), there will be no stopping them. The system is thus inherently unstable: it will expand at an exponential rate, only to come crashing down apocalyptically once all their possible prey in the city of New Crobuzon has been consumed. The proliferation of zombie bodies can only end in extermination and extinction.

Capitalism with an Inhuman Face. The slake-moths are cold-blooded, unromantic, and utterly inhuman. This is what makes them so different from Dracula and his kin. For we cannot empathize with the slake-moths or put ourselves in their place. We cannot imagine ourselves *becoming* them. The only relation we have to them is as prey. We are unable to resist when they seduce us with the "awesome, unfathomable, and terrible beauty" of their wings' visual display (528). But we cannot see ourselves in them when we watch them feed. First they emit "tiny, obscene noises" and "drool in vile anticipation" (363–65); then they grow "exhilarated" and "drunk" with "the entrancing, inebriating flow" of the psychic energy that they slurp up from their victims (647–48). This gibbering insect lust (as Burroughs might call it) is nothing of ours, nothing that we can recognize and own. The slake-moths are alien beings, creatures of sheer excess; this is how they embody the depredations of an inhuman vampire-capital. I don't mean to imply by this that the slake-moths are merely allegorical figures; the whole point of *Perdido Street Station* is to present them to us literally, as actual, material beings. But in their rapacity and insatiability—and indeed, in their very presence in the air over the city—the slake-moths are an expression, or better an *exudation* (to use a word Miéville favors), of the self-valorizing movements of capital. As psychic vampires who prey on imagination and thought, they enact the appropriation and accumulation of human mental creativity. They even transform this impalpable *intellectual capital* (or *intellectual property*, so-called) into a tangible commodity, in "quasi-physical form," by exuding a liquid that is "thick with distilled dreams" (377). This liquid provides sustenance to slake-moth caterpillars, which "cannot yet digest purely psychic food" (377). But it can also be harvested and processed into a potent psychedelic drug for human consumption, aptly known as dreamshit. To take dreamshit is to consume (or to be consumed by) "the barrage, the torrent, of psychic effluvia" of other minds (186). And indeed, the slake-moths are initially brought to New Crobuzon in order to extract and realize this psychic surplus value. The government uses them for mind-control experiments, and the mob milks them for drugs. For the authorities and their backers in big business, as for the criminal underworld, there is nothing wrong with the slake-moths themselves. There is only a problem when the creatures break free from their captivity, and start terrorizing the city indiscriminately. The escaped moths' unrestrained predation is a nightmare of surplus appropriation gone mad. The city's relations of production have started to metastasize and proliferate uncontrollably, like a cancer. But the crucial point about such a cancerous pathology is that it is nothing more than a monstrous intensification of the "normal" functioning of the system infected by it. The slake-moths do not represent an economy foreign to New Crobuzon; they are just capitalism with an (appropriately) inhuman face. They are literally unthinkable, yet at the same time, they are entirely immanent to the society that they ravage.

Dawn of the Dead. "What might zombies have to do with the implosion of neoliberal capitalism at the end of the twentieth century?" (Comaroff and Comaroff 17). In contrast to the inhumanity of vampire-capital, zombies present the "human face" of capitalist monstrosity. This is precisely because they are the dregs of humanity: the zombie is all that remains of "human nature," or even simply of a human scale, in the immense and unimaginably complex network economy. Where vampiric surplus-appropriation is unthinkable, because it exceeds our powers of representation, the zombie is conversely what *must be thought*: the shape that representation unavoidably takes, now that "information" has displaced "man" as the measure of all things. In a posthuman world, where the human form no longer serves as a universal equivalent, the figure of the zombie subsists as a *universal residue*. We all know those shuffling, idiot bodies, wandering obliviously through the country-side, wailing with a hunger that can never be sated, list-lessly tearing apart the flesh of the living. Jean and John Comaroff report that, in rural South Africa, "persons of con-spicuous wealth, especially new wealth whose source is neither visible nor readily explicable" are often accused of acquiring that wealth by "zombie-making" and exploiting zombie labor (20). The Comaroffs see this myth as a signi-fying displacement. Even as the commodity form triumphs universally, labor is hidden as the source of value. A capi-talism visibly defined by financial speculation and by the uninhibited flaunting of wealth comes to be "invested with salvific force" (19), as if it produced its riches by magical means. In the Third World, "millennial hope jostles material impossibility" (22); the evident affluence of the few coexists with massive unemployment among the many. Production has literally become invisible; it can only be imagined as an occult force. In such an atmosphere, the Comaroffs suggest, "zombie tales dramatize the strangeness of what has become real: in this instance, the problematic relation of work to the production of social being secured in time and place" (23–24). There is one additional twist to this scenario, however, that the Comaroffs fail to explore. The images of zombies in these stories are not derived from traditional South African folklore, but are rather taken full blown from American horror films. For there is no escaping the relentless circulation of images in the global mediasphere. Zombies originally came into American culture from distorted accounts of Haitian Voudoun. But now, the periphery indulges in a sort of reverse exoticism as it appropriates the mythologies of the imperial center. With one striking difference, however: in the American movies themselves, the zombies are not workers and producers, but figures of nonproductive expenditure. They squander and destroy wealth, rather than producing it. The classic instance of this is George Romero's *Dawn of the Dead*, in which the zombies converge on a huge indoor shopping mall because that is where they were happiest when they were alive. Even in death, they continue to enact the rituals of a rapacious, yet basically aimless, consumerism. As the zombie circulates between First World and Third, it also circulates between work and idleness, or alienated production and conspicuous consumption.

Snow Crash. Sometimes it gets to be too much. Sometimes you just want it all to stop. Deleuze and Guattari say it best: "Everything stops dead for a moment, everything freezes in place—and then the whole process will begin all over again. From a certain point of view it would be much better if nothing worked, if nothing functioned" (1983, 7). The network, like the galaxy, has a black hole at its center: a void into which all information disappears and out of which nothing ever returns. There's always a moment when the system freezes or crashes, and everything has to be rebooted from scratch. The disruption may come in the form of an information virus, as Neal Stephenson suggests in his novel *Snow Crash*. But it's just as likely to be a hardware device: the Information Bomb. A prototype I-Bomb has already been built by the Experimental Interaction Unit (EIU) in San Francisco. It releases "an extremely powerful electromagnetic pulse (EMP)" that "enables the disruption and/or disabling of electronic devices and systems within a limited range and without physical contact. It will also corrupt and/or erase any magnetic storage medium such as floppy disks, credit cards, etc." (Experimental Interaction Unit). So far, the I-Bomb has only been used in a limited series of tests. But I-Bomb attacks are common in the City of *Transmetropolitan* (Ellis and Robertson 2000, 31–34). The purpose of these terrorist actions is to create temporary Technology Free Zones (TFZs): spaces in which the absence of information devices will "allow direct human-human interaction to flourish" (Experimental Interaction Unit). In other words: "Talking. Holding hands. Telling stories. Dirty jokes. That sort of thing" (Ellis and Robertson 2000, 34). TFZs are Ellis and Robertson's post-network-society twist on Hakim Bey's idea of a Temporary Autonomous Zone (TAZ): "the TAZ is like an uprising which does not engage directly with the State, a guerilla operation which liberates an area (of land, of time, of imagination) and then dissolves itself to re-form elsewhere/elsewhen, before the State can crush it." The impermanence of the TAZ, its refusal to perpetuate itself by building stable, long-term institutions, is its most crucial feature. As for TFZs, they have their own built-in ironies, which also militate against judging them by the criteria of long-term success. For one thing, as the EIU manifesto itself points out, there's the irony that technology ("sophisticated mechanical and electrical systems") must be used in order to overturn technology. For another thing, TFZs are not really technology-free: after all, things like clothing and eyeglasses, which are not affected by the I-Bomb, are also products of human technology. The I-Bomb works better as a bratty, disruptive prank than it does as a total reinvention of humanity. Nonetheless, the experience of being in a TFZ is unexpectedly poignant and strange: "I've never known such silence in this City. The hum of wearable computers, the thump and distort of musics, the jabber of phones—all gone, suddenly. Eye of the storm" (Ellis and Robertson 2000, 33).

Bliss. For his part, K. W. Jeter is way too skeptical, or perhaps too misanthropic, to put much stock in such utopian spaces. As far as he is concerned, if you're in a crowd with other people, then you are still connected. At best, TAZs and TFZs might be momentary escapist consolations, much like McNihil's film noir visual overlay. But none of these contrivances actually gets away from the tyranny of the network. Even death, as we have seen, is not a true escape. In *Noir*, there is only one way out. It's a surgical operation known as the "Full Prince Charles" (FPC)—so called after the Prince of Wales, who told his lover that he'd like to be reincarnated as her tampon, "so that he could be with her forever, constantly in place where he most loved to be" (429). In an FPC, you shed most of your extremities and organs; you are "skinned and reduced, taken down to essentials, the other parts thrown away, trimmed into the scrap bin under the sink" (431). The operation is quite similar,

actually, to the punishment meted out to copyright violators; only the aim here is pleasure, rather than pain (431).
Instead of being put into a speaker cable or the electrical
cord of a toaster, you are inserted into a woman's uterus.
You're attached securely enough to the wall of the womb,
so that you won't be washed out of it again by the flows of
menstruation. After having an FPC, you are "no longer a
separate organism," since your "physiology is totally dependent upon the host"; you are nourished through a "permeable skin casing" (433). Though the Full Prince Charles
is largely a male fantasy thing (427), women as well occasionally have it done (431). For the psychological basis of
the FPC is less important than the fact that it is "the best
hiding place" anyone can find, "the ultimate, really" (428).
The interior of the womb is the one place where the debt
collectors and corporate stooges can't get at you. The only
way out is the way in and through. There is no place outside the reach of the network, but you can perhaps evade it
if you go deeply enough *inside*. The Full Prince Charles leads
to a sort of liminal consciousness, thought sinking down
toward the nothingness of fusion, but just distinct enough
to stop on the verge of its own extinction: "absorbed and
yet still separate. Or just separate enough to be conscious,
to know where you [are]..." (430). No sight (431) and no
hearing (434), just the barest glimmer and whisper of self-
awareness in the welcoming darkness and silence.

Switched Off. The silence only lasts for an instant, however. For there is no outside to the space of flows, no place
free from the relentless pressures of information. Modernity dreamed of Otherness and the Outside, but these terms
have no meaning in today's globally networked world. As
Hardt and Negri put it, "in the passage from modern to
postmodern and from imperialism to Empire there is progressively less distinction between inside and outside...

The modern dialectic of inside and outside has been replaced by a play of degrees and intensities, of hybridity and artificiality...The Other that might delimit a modern sovereign Self has become fractured and indistinct, and there is no longer an outside that can bound the place of sovereignty" (187–89). Of course, this does not mean that everyone now actually lives in McLuhan's global village, or Teilhard's noosphere. Indeed, for most people, across the globe, the overwhelming lived experience is one of poverty and deprivation. But this deprivation is no longer (if it ever was) the Outside of the dominant system, its limit and its threshold. For we no longer live under a regime of Malthusian scarcity. The network is a realm of pleonastic abundance. Everything is available immediately, from everywhere. In such a world, scarcity and deprivation are not "natural," given conditions; they are products of "hybridity and artificiality" (Hardt and Negri 188), conditions that must be deliberately introduced, produced, and reproduced. It is not that certain people and places aren't (or aren't yet) connected to the network. It is rather that these people and places have been actively *disconnected* from the network. As Castells says, "in some instances, some places may be switched off the network, their disconnection resulting in instant decline, and thus in economic, social, and physical deterioration" (2000b, 443). It is often argued that censoring the Internet is impossible; in the words of John Perry Barlow, when an obstacle to free expression arises, the network simply "reroutes proscribed ideas around it" (1994). But the network also reroutes itself around anyone who threatens its dominant power relations. People, and even entire regions, are made to disappear into what Castells calls "the *black holes of informational capitalism*" (2000a, 165). The metaphor is precise: if you fall into one of these black holes, you will never emerge out of it again. You are lost to the world where information is power. It used to be said that the poor were always with us. But in the network society, with the dismantling of welfare programs and the privatization of formerly public space, this is no longer the case. Instead, the poor are made to vanish; today, they are already gone from the purview of the average netizen or corporate manager. It only needs a few more steps to reach the situation in Jeter's *Noir*, where the rich can legally murder homeless people, or people without identity papers, as long as they pay some sort of compensation (for instance, by having their corporations plant seedlings along the highway, seedlings that will soon die anyway because of pollution), and preregister their killings with the police (61–77).

All That Is Solid Melts into Air. In *Noir*, Jeter shows us a future Seattle that has been turned into an informational black hole. There's an obvious irony here, since Seattle today is best known for its information technologies and software; Real Networks and Amazon.com are headquartered here, and Microsoft is just across the lake, in Redmond. An additional irony comes from the famous 1999 Seattle protests against the WTO and corporate-controlled globalization; the novel almost seems to prophesy these events, which took place a year after it was published. In Jeter's depiction, future Seattle is totally run down. It seems to consist mostly of porno movie theaters and derelict hotels where junkies watch TV all day long while receiving their drugs of choice through intravenous drips. There are also crowds of homeless people, who cannot even afford that level of comfort. Many of them live in a downed Boeing 747, whose wreckage is strewn all across downtown. This is yet another irony, for Boeing was the dominant corporation of the old, industrial, blue-collar Seattle, before the city was changed by the information economy. At a crucial point in *Noir*, the Boeing 747 bursts into flames, which are put out by a fire-extinguishing foam sprayed from helicopters.

An orgy ensues, as the 747's excited denizens wallow in the foam. Jeter describes it in page after page of delirious purple prose: "the incendiary rage had transmuted into a giddy frivolity, a damp carnival of billowing foam and slippery human skin. The smell of wet wood and other debris, floating on the soft whiteness as if it were a slow-motion sea, mingled with pheromone-laden sweat" (216). All that is solid liquefies, melting into water and sweat and sperm and vaginal fluid: "the strictures of form and identity dissolving, the prisoning matter of the city's heart reverting to some premammalian coitus. The way... that fish and things that swim around in the ocean do it" (217). This is the final dissolution of boundaries, "a horrifying *connectedness*" (218) so extreme that it far outstrips anything the network could do. The global flows of capital and information obliterate every obstacle in their path, trapping us in the network's sticky web. But at least, on the network, we are able to remain somewhat apart, in our separate nodes, each of us a monad without windows or doors. Communication and consumerism alike require at least a minimal distance between the connected parties, and a well-certified separate identity for each. Our network interfaces are like our skin and our other bodily membranes; they protect us from the outside world, even as they offer us a means of regulated exchange with it. Each membrane or interface is a *buffer*, in the double sense of something that cushions from shocks, and something that takes in, and stores, materials or messages. But in Jeter's vision of Seattle, there is no longer any buffer: "the distinction between one body and another was erased, the membrane between the body's interior and the soft outside world forgotten" (217). We come face-to-face with the catastrophe of absolute fusion, gravitational collapse into the singularity of a black hole. And Jeter wonders if this is indeed "the oncoming tide of the future, humans having finally gotten tired of bones and jobs to do" (217).

Fade Away and Radiate. Every catastrophe demands a witness. There must be a Berkeleian observer who is unaffected by the disaster, and in whose ideas or perceptions it thereby becomes real. *And I only am escaped alone to tell thee.* The physicists tell us that even a black hole is not entirely opaque. It may swallow up unlimited quantities of matter and information, but it gives them back in the form of Hawking radiation. Particle/anti-particle pairs arise spontaneously out of the void; usually the particle and the anti-particle annihilate each other right away and return to nothingness. But if the pair is generated close enough to the event horizon of a black hole, it can happen that one of the particles is drawn into the singularity, while its counterpart escapes and radiates outward. Much the same thing holds for informational black holes. These people may have been thrown out of the network, but they are not altogether beyond its reach. They do not participate in the global circuits of production and exchange, but they still radiate residual energy back into those circuits. This energy gets picked up in the form of relentless media coverage. Political protestors have long had to deal with the irony that, no matter what they do, the revolution *will* be televised. The televisual event horizon was passed sometime in the 1960s. Ever since then, in the words of Brian O'Blivion, the McLuhanesque media sage of Cronenberg's *Videodrome,* "whatever appears in the television screen emerges as raw experience for those who watch it. Therefore, television is reality, and reality is less than television." Time and again, activists have invented all sorts of creative and subversive ways to manipulate the very media that have been manipulating them. But they have never been able to erase the fact that, precisely because of ubiquitous mediatization, their resistance is always compromised from the outset. This is why Jean Baudrillard (1983) scorns political activists for their inadvertent complicity with the system that they claim to oppose. On the other hand, Baudrillard romanticizes the "silent majorities," whose passive indifference is supposed to exempt them from the machinations of power.

The Whole World Is Watching. Jeter is fascinated, but also filled with disgust, by the insurmountable passivity of the masses. The orgy in Seattle is the *ne plus ultra* of Baudrillardian indifference; here the masses indeed simply ignore power, rather than trying in any way to resist it. Swept away in a collective frenzy, the revelers merge into "one compound animal, a colony of undifferentiated sensual function" (396). The authorities are unable to control the riot, precisely because it isn't in any sense directed against them. Instead, they choose to contain it, by encasing downtown Seattle in a semitransparent gel, a "sterile nutrient medium" (392) that isolates the rioters while keeping them alive. The "partially dissolved once-were-humans" do not ever seem to notice. They continue to thrash about tumultuously, or to drift "on slow currents through the gel, mingling their soft bones and loose organs with each other in lazy pre- and postcoital suspension," their skin chafed away, their bones dissolved, their extruded neurons "hooked up and knitted together" like seaweed (398). For Jeter, this spectacle of dissolution and fusion is the hideous form of fully "requited desire...that's what you get, when you finally get what you want...given half a chance, people would slough off the soft, thin barriers between themselves and achieve a nakedness of the exposed flesh, perfect for non-stop connecting" (397–98). But this "poly-orgynism," this "ultimate connection" (395) beyond the network, is also an object of media fascination, every bit as much as Seattle's 1999 WTO protest was. All the TV networks provide continuous coverage; the "sheer commercial appeal" of such "real-time pornumentaries" (395) is immense. This is how the catastrophe finds its witness. The informational black hole provides entertainment for the very society that disconnected it and excluded it as waste. If the network is full of desperate exhibitionists, each of them presenting the spectacle of his or her life in a Web cam broadcast that nobody else watches, then the life of the silent masses is precisely the reverse: collectively, they present a spectacle that is watched by everyone else but of which they themselves remain blissfully unaware.

Red Spider, White Web. Where Ballard's *Super-Cannes* explores the space of flows from the viewpoint of the ruling elites, Misha's science fiction novel *Red Spider, White Web* maps this space from the viewpoint of its victims. The novel is set mostly in the aptly named Ded Tek, a postindustrial wasteland of abandoned factories, empty warehouses, disused railway yards, and polluted rivers. These are the vestiges of the old manufacturing society, like "the fabulous ruins of Detroit."[70] They were left to rot after the triumph of the virtual "new economy." But Ded Tek is not empty, far from it. It burgeons with all sorts of hideous life, such as the "long and repulsively turd-like" mutant trout that Misha describes at one point: "[a]slimy brown thing... like a legged lungfish, only big as a small sturgeon, and glistening with oil-slick colors. It had a huge mouth, which was barbed with catfish whiskers, and it was covered with orange, oozing ulcers" (97–98). The artists who are the novel's main characters are the human inhabitants of Ded Tek. They are not much better off than that trout. They must struggle day by day to eke out a meager living, menaced by flesh-eating zombies, fanatical religious cults, serial killers, cops, and other dangers. These artists have all been excluded from the network for failing to be economically productive. The last joints of their right index fingers have been lopped off as a legal mark of their offline status. But they are never free from the network economy, even though they cannot participate in it. No matter what they do, it still exploits them. For nothing is outside the virtual space of flows. The artists' own work is virtual, for one thing. It consists of holographic displays: beautiful three-dimensional scenes, unfolding in continual metamorphoses, with full color and sound. For another thing, the artists have to sell their work somehow, simply to survive. Whether they find patrons to buy their pieces or sell out for the greater rewards (but also greater humiliations) of wage labor, they have to make compromises of some sort. Inevitably, they end up being targets of abuse. Only instead of facing executives out on a therapeutic *ratissage*, they encounter the teenaged children of the elite, futuristic frat boys more or less, who go around Ded Tek in gangs, beating up artists and other transients just for the hell of it: "rich kids, scrubbed so clean they stank, wanting to get off by splashing a bit of filth on the tips of their soapy dongs" (31).

Living Metal. Ded Tek is a landscape of "empty warehouses and strange, rusted towers," with "a debris of waste papers, glittering glass, windblown, dry, grass and broken hunks of asphalt" scattered underfoot, permeated by slowly oozing chemical poisons (14). The old industrial machinery no longer works the way it was originally designed to. Rather, its gears, cables, and diodes have been twisted into ungainly shapes by the forces of decay, or scavenged by artists and repurposed for monstrous new uses. Everything in *Red Spider, White Web* is caught in an inhuman, slow-motion metamorphosis. Ded Tek is "a world of living metal" (12–13), full of what Deleuze and Guattari call non-organic life: "what metal and metallurgy bring to light is a life proper to matter, a vital state of matter as such, a material vitalism . . . Not everything is metal, but metal is everywhere. Metal is the conductor of all matter . . . Thought is born more from metal than from stone" (1987, 411). Indeed, Misha's own prose could be called metallic: "The echoes of voices, laughter, cursing, dropped metal and shouts all slammed into Kumo's ears and she shuddered in revulsion. The old smell of unwashed bodies, of clone leathers, of cigarettes and synthetic sushi clammed onto her senses like a wet blanket" (39). In such sentences, words are like blows; they clang together harshly. The tongue stumbles over the densely clotted consonants. Misha's language is densely paratactic; it is not organized by any hierarchical logic, but only by subliminal rhymes (slammed/clammed) and a sort of stuttering, continually interrupted rhythm. In this way, Misha's prose invokes (as well as describes) a world of intense materiality. Sensations are not perceived at a disinterested aesthetic distance; rather, they assault the protagonist Kumo with a deadly, visceral force. But conversely, and at the same time, it is only out of such violent experiences in the "world of living metal" that Kumo and the other artists of Ded Tek are able to create their holograms, all-encompassing virtual illusions of an intoxicating beauty. In this way, the wasteland of Ded Tek is a place where the virtual world encounters the physical one, and each is transformed by, and into, the other. The crass material weight of the older, dead technology is alone what makes possible the magical, airy lightness of the new. The wealthy inhabitants of the space of flows regard art and artists, at best, with a patronizing condescension; more often, with contempt. They exclude art from their everyday lives, just as they do any other sort of waste. But of course, since they live in a wholly simulated environment, everything that surrounds them in fact is art. Their dirty secret is that they can only live in such virtual splendor by appropriating the very thing they scornfully reject. The art they scorn is their ultimate source of value. Everything in the world of *Red Spider, White Web* is a virtual construct, built from the detritus and dead labor of Ded Tek.

The Day the World Turned Day-Glo. "Candyland. Lollipop lights, syrupy buildings, gumdrop houses, sugar-frosted fake turf, rock-candy gardens, koi pools, ginger bridges. Everything was lit from inside like a garden of edible jewels...The light twinkled through lilac, glowed through pink, and shimmered through green" (Misha 199). This is the city of Mickey-san, home to the privileged, idle rich in *Red Spider, White Web*. As its name suggests, Mickey-san is a transnational hybrid: a cross between a fun-filled Japanese theme park and a Disney-engineered, family-values-oriented "community of tomorrow."[71] It's a gated development, where you can live free from stress and worry, with all the physical comforts and virtual distractions you want. There are talking animals everywhere and soothing music emanates from the walls. Safety reminders, such as no-smoking messages, pop up before you have even have a chance to get yourself into trouble. Most people in Mickey-san wear "self-communicative suits" (204), narcissistically reinforcing virtual-reality gear. Everything is refracted through a "kaleidoscope of color, so wildly different from the galvanized and concrete world of Ded Tek" (199). Mickey-san is the last word in hyperreality: a living cartoon, a holographic wonderland. And Misha delineates, with deadpan hilarity, what hyperreality *feels like*. It's a diffuse, soft-edged psychedelic sensation, more like Ecstasy than LSD. After all, don't Ecstasy and Disney World have at least this one thing in common, that they both make you feel like everyone loves you and that you love them all back? In

Mickey-san, the prevailing mood is always relentlessly up-beat. Physical objects are more vivid than in ordinary life, but also somewhat blurred around the edges. Sensations are heightened: not brutally, as they would be in Ded Tek, but with an all-suffusing, forcibly cheery warmth. In Mickey-san, you forget about the past and the future and live solely in an eternal present. The inhabitants don't produce anything themselves; as the disgruntled artist Mottler sees it, "all they did was consume... their synapses were so fried from the suits, he doubted they had anything more than short-term memory" (208–9). Is it merely a coincidence that a similar pattern of synaptic damage is alleged to result from the long-term use of Ecstasy? Or are these yet more examples, together with optimized transience disorientation, semiotic AIDS, and I-Pollen Related Cognition Damage, of a pathology intrinsic to information society?

Remix/Remodel. Drugs are a kind of technology, just like clothing, or the wheel, or telephones, or computers, or indoor plumbing. They are *media,* in McLuhan's sense of the term: "extensions of ourselves," prosthetic amplifications of our bodily powers. Indeed, when McLuhan talks about the media, it often sounds as if he is actually describing drugs: "any invention or technology is an extension or self-amputation of our physical bodies, and such extension also demands new ratios or new equilibriums among the other organs and extensions of the body" (1994, 45). Whenever we worry about drugs altering our bodies and minds, we should remember that cars and television do this too. The difference is really just one of scale. High-speed travel and long-distance instantaneous communication are experiences that have reorganized our senses, over the course of the last two centuries, in a massive, macroscopic way.[72] In contrast, drugs affect our bodies and minds on a far more intimate—microscopic and molecular—level. Drugs work as they do because they are close analogues of neurotransmitters that are already active in our brains. They work, therefore, as agonists or antagonists of the various neurotransmitter receptors. Opiates mimic the effect of endorphins; LSD and other psychedelics bind to the serotonin receptors (as do antidepressants); cocaine stimulates dopamine release by inhibiting its reuptake.[73] Taking drugs is thus a kind of biological engineering. In aesthetic terms, it is a way of experimentally *remixing* consciousness, much as a DJ remixes musical recordings.[74] The *scandal* of drug use comes largely from the fact that, in contrast to a DJ or a research scientist, one is experimenting primarily upon oneself. It is often objected that drugs are not safe, but isn't that precisely the reason for taking them? With psychedelic drugs especially, self-experimentation that violates the traditional norms of scientific objectivity is structurally necessary to the actual research, as Richard Doyle (2002) argues. This is because the effect of these drugs is to transform irreversibly the very subject who has been examining them. Taking LSD is a jump into the unknown. I am no longer the same person I was before the experiment began.

The Antinomy of Psychedelic Reason. The psychedelic experience short-circuits the distinction between the observing self (the self as transcendental subject, or subject of enunciation) and the observed self (the self as empirical entity, or subject of statements). However, this double structure, that of Lacan's "split subject" (138ff), or Foucault's "empirico-transcendental doublet" (1970, 318ff), is central to the very project of scientific reason. Its collapse, under the influence of LSD and other psychedelic drugs, means a radical mutation in the space of our understanding. Psychedelic experience compels us, on the one hand, to concede more to the biochemical realm than we might otherwise want to (since these drugs wreak havoc upon our ideas of free will and responsibility) and, on the other hand, to reject any sort of biological determinism (since these drugs teach us how to induce biochemical changes at will—even if we are not in control of the consequences). In this way, the psychedelic experience lies at the heart of the Kantian antinomy that structures our contemporary understanding of consciousness. On the one hand, we reduce mind to a mere series of biochemical effects; on the other hand, we inflate it into a principle of authenticity, or even (thanks to New Age spirituality) into a transcendent and universal ground. The psychedelic experience unavoidably gives rise to both of these illusions of reason.[75] As Erik Davis says, "psychopharmacology is simply offering its own resolutely philosophical answer to the eternal problem of human suffering: Use technology to control its symptoms . . . The paradox is that these mechanistic molecules can produce deeper, more authentic selves." When you're tripping, it's hard not to be amazed that all this is due to a simple chemical change in the brain. But it's equally hard not to be seduced into thinking that you have somehow attained spiritual enlightenment, or found the essence of the self, or gained access to "different levels of reality . . . other planes of existence" (Strassman 315). This paradox can be resolved, however, in a classically Kantian manner. The reductive, instrumentalist view of the mind on drugs comes from seeing the self as just another empirical object to be tweaked by engineering like all the rest, while the expansive, metaphysical view of the mind on drugs comes from an unwarranted absolutization of the self's position as transcendental subject. The two sides of the antinomy are drawn from the two sides of the "empirico-transcendental doublet," and both are undone, simultaneously, by the same movement that initially generated them: the unpredictable metamorphoses of the subject of psychedelic experience. This nonself-identity is an enigmatic, irreducible limit, both beckoning thought and repelling it.[76]

Homage to Psychedelia. Psychedelic drugs and electronic technologies affect the sensorium in strikingly similar ways. They both disperse and decenter subjectivity. Consciousness is scattered all across space, and yet in a strange way intensified. Concentration is no longer possible, for too many things are happening all at once. Forms are fragmented, and context is erased. Each event seems to stand by itself, throbbing with its own singular halo. Objects turn into hyperreal doubles of themselves: dense clouds of information, pulsing with bright, saturated colors. Yet, despite the isolation and fragmentation, everything seems to be mysteriously connected. You are in direct communication with the most distant forces of the cosmos. You may feel disengaged from your immediate physical body, but you are caught up instead in a new process of virtual, prosthetic re-embodiment. In all these ways, psychedelics are the drugs that resonate most powerfully with the space of flows. It's not surprising that the discovery (in 1943) and dissemination (in the 1950s and 1960s) of LSD roughly coincide with the development of the digital computer. And as Richard Doyle (2002) suggests, there was also considerable overlap between LSD research and the elaboration of the technologies of molecular genetics, based on the manipulation of DNA. If psychedelic drugs are indeed leading us, as the late Terence McKenna (1992) claimed, into an era of "archaic revival," then this is the case for good McLuhanesque reasons. The culture of industrial modernity was visual, linear, rationalized, and homogeneous. But postmodern network culture, just like many so-called "primitive," preliterate cultures, is audio-tactile, an allover field of heterogeneous connections. Today, once again, we are

living "in acoustic space: boundless, directionless, horizon-
less, in the dark of the mind, in the world of emotion, by
primordial intuition, by terror" (McLuhan and Fiore 1967,
48). Such were the contours of the "archaic" world of magic
and spirits, but they are also those of the "world space of
multinational capital" today. It's a space that, as Fredric
Jameson insists, is strictly speaking "unrepresentable,"
though not necessarily, for all that, "unknowable" (53–54).
This distinction is a crucial one.[77] Representation cannot comprehend the network, just as it cannot encompass the psychedelic experience. Indeed, this is why verbal accounts of
"trips" are generally so lame and inadequate.[78] The "terror" of acoustic or network space, like that of psychedelia,
is the flip side of an equally boundless exhilaration. LSD
shakes consciousness to its foundations; on the way to remaking our perceptions of the real, it blocks our usual
schemas of mental representation. But precisely by over-
flowing the limits of representation, LSD mimics, and coincides with, the flows of the network itself. LSD doesn't "represent" the space of multinational capital, but it doubles
that space, accompanying each of its folds and convolutions and miming each of its convulsions. Despite its shattering intensity, LSD remains a drug of surfaces and mere
appearances. It is always—if ever so slightly—more virtual, more artificial, and more superficial than the space of
flows itself. And that is what gives it such a privileged role
in the network society.

Before and after Science. We should not exaggerate
the similarities between postmodern and "archaic" cultures,
however. For they are as profoundly different from one another as synthetic chemicals like LSD are from mind-
enhancing substances found in plants (ayahuasca) and fungi
(psilocybin). McKenna insists upon this distinction. LSD is
to organic psychedelics, he says, as a flight in a Sopwith

Camel (in which "you could hear the air shrieking over the control surfaces and feel the wind blasting your face") is to a voyage on the space shuttle (Gehr). This is a strange and interesting analogy. Although McKenna values organic substances and disparages the artificial products of the laboratory, he describes the former, no less than the latter, in terms of technology. Moreover, he depicts the laboratory product as a clunky, underdeveloped technology that fails to get very far off the ground (a World War I biplane), while he conceives the organic substance as almost like a perfect simulation, a virtual-reality ride providing a seamless experience of cosmic scope (viewing the earth from orbit). McKenna himself compares the chattering elves he encounters while tripping on DMT (the active ingredient in ayahuasca) to the Munchkins in *The Wizard of Oz:* "the tryptamine Munchkins come, these hyperdimensional machine-elf entities, and they bathe one in love. It's not erotic but it is open-hearted. It certainly feels good" (1983). But why does McKenna's sacred "hyperspace" sound so much like Baudrillard's (or Disney's) hyperreality? It shouldn't be this way, given that McKenna regards DMT and psilocybin as "technologies of the sacred," rather than as mere "extensions of man." That is to say, these technologies are imbued with otherness. They are not just alienated, "autoamputated," portions of ourselves. They come from someplace other than the human body and mind. It follows that *human beings are the object of these technologies, but not their subject.* McKenna thus offers a more benign version of Burroughs's insight that language is a virus. Thanks to DMT, McKenna says, "one hears and beholds a language of alien meaning that is conveying alien information that cannot be Englished" (1983). And this, I think, is both the strength of McKenna's insight and the point where the problem comes in. What compromises McKenna's otherwise radical formulation is his claim that the alien language is nonetheless still conveying "information."[79] Which means that, even if this alien language "cannot be Englished," it can be binarized and digitized. McKenna forgets that "archaic" society has no concept of information, just as it has no concept of money. It is only in capitalist modernity that money, and subsequently information, are unleashed to become universal equivalents. Once McKenna admits the notion of information, his insights about radical otherness are lost. Soon he is speaking, not of an alien language, but rather of "the assembly language that lies behind language, or a primal language . . . a primal 'ur sprach' that comes out of oneself" (1983). And instead of dispersing and shattering the ego, DMT now appears to be rebuilding it on a transcendent plane, by "releasing the structured ego into the Overself" (McKenna 1984). My point is not the easy deconstructionist one that McKenna is metaphysically deluded, but rather the historicist one that, in our electronically networked society, the "archaic revival" can only take the form of an audioanimatronic simulation. *It's a small world, after all . . .* That is why LSD, with all its inauthenticity and artifice, is a more appropriate substance for today than McKenna's beloved organics. LSD is the right drug for our world, in which even delirium is a commodity, and "nature" no longer exists. Rather than trying to employ psychedelic drugs as spiritual tools, we should be attentive to their profane uses. We don't need shamans who visit the world of the dead or establish contact with the primordial forces of Gaia, so much as we need ones who can ride the financial flows and trace the global metamorphoses of capital.

Contentless and Abrasive. McKenna (interviewed in Hayes 417) says that LSD is "psychologically abrasive, like pouring Drano into your psyche." Where DMT seems to put us in contact with hyperdimensional entities, and psilocybin "has a message" it wants to tell us, LSD "is somewhat contentless." It doesn't actually make anything happen; it doesn't transport us to alien places, and it doesn't even tell us anything new. It just magnifies and intensifies whatever is already there, in your mind (the "set") or your surroundings (the "setting"). LSD is contentless and abrasive in that it strips away all determinate qualities and all particularities of content, transforming them into sheer, quantitative intensity—what Kant (1987, 103–17) calls the Mathematically Sublime. We can thus say of LSD what Marx and Engels famously said of capital: "All fixed, fast frozen relations, with their train of ancient and venerable prejudices and opinions, are swept away, all new-formed ones become antiquated before they can ossify. All that is solid melts into air." LSD casts itself over the entire range of phenomenological experience. It is a universal equivalent and translator for affect, just as money is for commodities, as DNA is for living bodies, and as digitized information is for all the media of expression. Dollars, bytes, DNA, and LSD: these are the magical substances, the alchemical elixirs, of global network culture. In all these cases, a "universal equivalent" imposes itself upon, and homogenizes, what was previously a heterogeneous group of materials. The equivalent first works as a common measure, or a simple medium of exchange, but soon enough it outgrows its utilitarian function and takes on a strange new life of its own. It ceases to "represent" the objects against which it is exchanged, and instead appears as their ground, or their animating principle. Money is no longer just the measure of wealth, for instance, but its origin. Commodities are not exchanged by means of money, so much as money realizes itself by means of commodities. This process is precisely what Marx calls the fetishism of commodities. More generally, it is the production of what Deleuze calls a *quasi-cause*. Every event, Deleuze says, "is subject to a double causality, referring on one hand to mixtures of bodies which are its cause and, on the other, to other events which are its quasi-cause" (1990, 94). The actual cause of any event is a physical "mixture" that remains external to the event itself. (For instance, the intensified exchange of neurotransmitter chemicals at certain synapses in the brain is the actual, material cause of LSD hallucinations.) But the quasi-cause is a virtual, incorporeal process, immanent to the event that is its effect. (To keep the same example, a vibrant upswelling of cosmic forces is the virtual cause, or quasi-cause, of LSD hallucinations.) The quasi-cause is not an actual, empirical phenomenon, but it should not be dismissed as a mere illusion. Finance, information, and psychedelia are no less real for being virtual; indeed, the fact that we are unable to place them in any particular physical location is one reason for their uncanny power.

Me and My Shadow. In postmodern society, everything flows. There are flows of commodities, flows of expression, flows of embodiment, and flows of affect. The organizing material of each flow is a universal equivalent: money, information, DNA, or LSD. But how are these flows related among themselves? Strictly speaking, they should be interchangeable. All the equivalents should themselves be mutually equivalent. Now that everything has been subsumed under capital, now that everything is a commodity, it is no longer possible (if it ever was) to divide things between an economic base and a noneconomic superstructure, as the older Marxist theory used to do.[80] Today, culture and politics coincide with the economic sphere; they are themselves, directly and immediately, processes of production. In the language of Spinoza: money and information, and the other universal equivalents, are immanent attributes of one and the same social substance. And, as in Spinoza's philosophy, there is a strict ontological parallelism among these attributes. As Deleuze explains, "no attribute is superior to another... [the various attributes] present not only the same order, but the same chain of connections under equal principles" (1988b, 88). Ubiquitous commodification and ubiquitous informatization are the identical process, viewed under different aspects. In the theory of parallelism, therefore, there is no room for representation. Thought does not represent extension, but is wholly equivalent to it. Culture does not represent economic forces, but coincides with them. (It's not the ideology of Donald Duck that we have to worry about, so much as his trademark and copyright status.) Nonetheless, within Spinoza's theory, thought still plays a special role. Spinoza credits thought with the unique ability to *multiply* its objects or ideas, to *redouble* them, and thereby to *comprehend* them (Deleuze 1988b, 86–91). That is to say, thought's special privilege comes, not from any priority it might be expected to have, but precisely from its secondariness: the fact that it is repetitive and reflexive. The flow of thought, which is also to say of affect,[81] parallels the flows of all the other attributes, but with an infinitesimal delay, like an echo. And this gap, this echo or shadow, this time and space of doubling, is where subjectivity is born. Postmodern social space is traversed by financial, informational, and biological flows; the flows of affect modulate and shadow these other flows, repeating them in the mode of felt experience.[82] And this is precisely the function of psychedelia: as Deleuze and Guattari say, "the basic phenomenon of hallucination *(I see, I hear)* and the basic phenomenon of delirium *(I think...)* presuppose an *I feel* on an even deeper level, which gives hallucinations their object and thought delirium its content" (1983, 18). *Sentio,* rather than *cogito,* is the psychedelic first principle of postmodern consciousness.

Babylon Babies. What mutations, what transformations, might emerge from the space of flows? And how might psychedelic consciousness facilitate these transformations? Maurice G. Dantec's science fiction novel *Babylon Babies* explores these questions. The novel envisions the birth of a new posthuman species, *Homo sapiens neuromatrix*. Cloning, neurochemistry, shamanism, and artificial intelligence converge to create a mind of planetary scope. But Dantec's vision of the Singularity has little in common with those of Vinge, Kurzweil, and Moravec. For one thing, the path to Dantec's future is far from linear, in contrast to the way that Kurzweil and Moravec conceive the relentless march of technological progress: "history is an illusion; there is [only] process ... history is not pre-written and does not follow a causal logic, because there are no rules other than those of deterministic chaos" (558). For another thing, where Kurzweil tends to minimize the importance of the body, and Moravec to dismiss it outright, Dantec insists upon embodiment: "artificial intelligence consists precisely in inventing new types of corporeality. No intelligence can do without a body, without flesh; no artificial mind will ever emerge from simple numerical calculations, in a purely abstract space, the putrid old idealist dream of Hegel, Plato, all those old farts" (514). Dantec's "process" is a highly contingent one, involving the conjunction of all sorts of flows: flows of materials, flows of information, flows of consciousness, flows of weather, and flows of social and political conflict, as well as flows of biotechnology and computer technology in the strict sense. The unanticipated crossing of all these flows leads to the process of posthuman mutation. And in turn, the consequences of this mutation cannot be controlled, or even accurately predicted. There is nothing to do except to nurture the process, to watch over it, and to remove the obstacles in its path. Dantec's hackers and shamans take it upon themselves, as their ethical duty, to be loyal to the open possibilities of the future, instead of seeking to preserve the past: "we are not working for anybody,

said Lotus, but only *for the accomplishment of the process"*
(502–3).

Embrace the Serpent. As *Babylon Babies* tells its story
of the emergence of the posthuman, it also explores postmodern space. On one hand, it is the delirious narration of
a nonlinear process; on the other, the tortuous construction
of a phantasmatic cartography of flows. Of course, these
two projects are closely related. Marie Zorn, the young
schizophrenic who gives birth to posthuman infants at the
end of the book, learns how to "channel" the "uncontrollable mental flood" of her madness. She cannot limit this
"flood," or choose its contents, but she can narrate it and
thereby "make something of it" (115). Marie has also encountered—that is to say, her brain has established an interface with—a *neuromatrix:* "a network of artificial neurons
grown on DNA biofibers, and connected to electronic input/
output devices that served as its organs of perception"
(152). Through this link, she has access to all and any information in the world network. On top of all this, Marie also
provides a surrogate womb for illegal clones the Russian
mafia hopes to deliver to a North American New Age cult.
But soon, the deviant neurotransmitters coursing through
Marie's brain and body interact with the clones' DNA, trig-
gering mutations in the fetuses, which in turn feed back
into Marie's delirium. The fortuitous confluence of these
events is what sets the posthuman *process* into motion. The
self-narration provided by Marie's "linguistic schizo-
processor" (171) helps to organize the novel's own narrative. And the many branchings and connections, through
which Marie makes contact with the most heterogeneous
and most dispersed forces all across the world, is the
framework for the novel's own mapping of the space of
flows: "a schizo can perfectly well live having one part of
her body in Moscow, and the other in Ushuaia, or even on

Ganymede" (558–59). *Schizoanalytic cartographies,* in the phrase of Guattari: it's a matter of knowing how to follow the folds and convolutions of the space of flows, how to ride its turbulences, how to surf through its intersections and slide along its equivalences. The hackers of *Babylon Babies* use psychedelic drugs to follow what comes "naturally" (559) to the schizophrenic: to make connections between human consciousness and artificial intelligence, between individual subjectivity and global flows of information, and between the inner, twin-coiled "Cosmic Serpent" of DNA (552—referring to Narby) and the outer world of pheno-typic expression. This is how LSD mimes, doubles, and stands behind the flows of money, biomass, and informa-tion that pervade the network society.

Cocaine. In certain ways, it might seem that cocaine, rather than LSD, is the symptomatic drug of the world-wide marketplace. Cocaine certainly has a much greater role in the world economy than do any of the psychedelic drugs. If LSD follows the global flows of capital, cocaine simulates (and stimulates) the minds of the ruling elite, the people who actually live off of those flows. Just listen to one user's testimony: "I can actually feel the drug's power on almost a daily basis. I feel a deep surge of energy. It is less edgy than a double espresso, but just as powerful. My attention span shortens. I find it harder to concentrate on writing and feel the need to exercise more. My wit is quicker, my mind faster, but my judgment is more impulsive. It is not unlike the kind of rush I get before talking in front of a large audience, or going on a first date, or getting on an airplane, but it suffuses me in a less abrupt and more con-sistent way. In a word, I feel braced. For what? It scarcely seems to matter. And then, as the cocaine peaks and starts to decline, the feeling alters a little. I find myself less re-served than usual, and more garrulous. The same energy is there, but it seems less directed toward action than toward interaction, less toward pride than toward lust. The odd thing is that, however much experience I have with it, this lust peak still takes me unawares. It is not like feeling hungry, a feeling you recognize and satiate. It creeps up on you. It is only a few days later that I look back and realize that I spent hours of the recent past socializing in a bar or checking out every potential date who came vaguely over my horizon. You realize more acutely than before that lust is a chemical. It comes; it goes. It waxes; it wanes. You are not helpless in front of it, but you are certainly not fully in control. Then there's anger. I have always tended to bury or redirect my rage. I once thought this an inescapable part of my personality. It turns out I was wrong. Late last year, shortly after a snort of C, my dog ran off the leash to forage for a chicken bone left in my local park. The more I chased her, the more she ran. By the time I retrieved her, the bone had been consumed, and I gave her a sharp tap on her rear end. 'Don't smack your dog!' yelled a burly guy a few yards away. What I found myself yelling back at him is not printable in this magazine, but I have never used that language in public before, let alone bellow it at the top of my voice. He shouted back, and within seconds I was actually close to hitting him. He backed down and slunk off. I strutted home, chest puffed up, contrite beagle dragged sheepishly behind me. It wasn't until half an hour later that I realized I had been a complete jerk and had nearly gotten into the first public brawl of my life. I vowed to snort cocaine at night in the future."

A Syringe Full of Manhood. I must confess that I have been cheating a little. For the words in the above paragraph are not actually those of a cocaine addict. Rather, they come from Andrew Sullivan, in *The New York Times Magazine*, rhapsodizing over the wonders of testosterone. I have altered Sullivan's text only slightly: by changing the name of the drug, a few details about how he takes it (since testosterone is injected, whereas cocaine is snorted), and how often (since testosterone is taken, and has its effects, on a biweekly basis, while cocaine has rapid onset and short-lived effects). Sullivan has a medical prescription for testosterone. He is HIV-positive, as a result of which, he says, a few years ago, his body "was producing far less testosterone than it should have been at my age." To compensate, he injects himself every other week with what he calls "a syringe full of manhood." The result is a Jekyll-Hyde transformation, similar to what happens to users of cocaine, methamphetamines, and other chemical stimulants. There's the same rush of excitement, the same megalomania and sense of omnipotence, the same messianic zeal about the drug's miraculous effects, and the same fatuous, overbearing urge to describe the experience to everybody, in excruciating detail. There's also the difficulty that heavy drug users frequently have in expressing themselves clearly. Sullivan admits at one point that "I missed one deadline on this article because it came three days after a testosterone shot and I couldn't bring myself to sit still long enough." Later, he attributes any bias in his article to the fact that he wrote it "two days after the injection of another 200 milligrams of testosterone into my bloodstream." But he concludes that all these problems with the drug are unimportant compared to "how great it makes me feel." For Sullivan, the testosterone coursing through his veins is proof that men are fundamentally different from women. Women simply do not have enough of "the big T"; hence, they cannot be expected to have the "energy, self-confidence, competitiveness, tenacity, strength and sexual drive" that men do. Above all, women do not have that masculine "ability to risk for good and bad; to act, to strut, to dare, to seize." (It is unclear whether Sullivan would make an exception for such seemingly testosterone-fueled women as Ayn Rand and Margaret Thatcher.) The article ends, as drug rants all too often do, with a grandiose, self-congratulatory declaration. Sullivan tells us how "perfectly happy" he is "to be a man, to feel things no woman will ever feel to the degree that I feel them, to experience the world in a way no woman ever has." In short, the wonder drug testosterone has made Sullivan into a man, and a risk-taking, entrepreneurial man at that.[83]

Sense and Antisense. The delirium of LSD and other psychedelics marks the outer limit of the space of flows, its point of maximum expansion. Conversely, the delirium of testosterone, or of cocaine, marks its inner limit, its point of maximum contraction. On the one side, an ecstatic, self-dissolving embrace of the cosmos and all its flows, a swooning abandonment that has all too often been stereotyped as feminine; on the other, a will to master the flows (or to die trying) in the name of masculine self-affirmation. Thus Sullivan celebrates "a certain kind of male glory...valor or boldness or impulsive romanticism...a uniquely male kind of doom." At one inadvertently comic moment, he even worries that, if testosterone supplies were depleted, not only would straight men be domesticated forthwith, but "even gay men might start behaving like lesbians, fleeing the gym and marrying for life." In contrast, McKenna (1991) calls for the "breakup of the pattern of male dominance and hierarchy"; he celebrates "the feminizing of culture" and the cult of "the Great Goddess." Unfortunately, yet all too predictably, there isn't much difference between McKenna's positive vision of a transcendent, nurturing Mother Gaia, and Sullivan's disparaging description of "low testosterone" femininity as a state of "patience, risk aversion, empathy." For both authors, women are not subjects. They are only conceived as Others, as when McKenna praises "the presence of the Other as a female companion to the human navigation of history." Yet again, woman plays nature to man as culture. It little matters that McKenna urges "us" to welcome this feminine presence, whereas Sullivan warns "us" against it. But my aim here is not just to point out the deficiencies of McKenna's, as well as Sullivan's, gender politics. I also want to suggest how postmodern subjectivity may be understood in terms of the drugs that speak through these two authors. "Feminizing" psychedelics heighten and broaden consciousness, spreading it far beyond the confines of the ego, while "masculinizing" stimulants like cocaine and testosterone sharpen it, focusing it narrowly upon the body. Postmodern subjectivity moves back and forth between these extremes, as between the corresponding gender stereotypes. But with both extremes, subjectivity is compromised, even as it is being potentiated. For whether I take on the "feminine" role or the "masculine" one, whether I give myself over to the flows, or try to appropriate them to myself, I still must acknowledge that they were here first, and that my very being is predicated upon them.

Vegetable Mind. McKenna urges us to "adopt the plant as the organizational model for life in the twenty-first century, just as the computer seems to be the dominant mental/social model of the late twentieth century and the steam engine was the guiding image of the nineteenth century" (1991). Western thought is traditionally "based on animal organization," but the psychedelic turn could lead instead to "our reinvolvement with and the emergence of the vegetable mind." Plants are exemplary, for McKenna, both because of the psychedelic substances they provide and because they are self-maintaining organisms, able to manufacture their own sustenance. Vegetable mind might seem to be an oxymoron, since plants do not have neurons, nervous systems, or brains. But surprisingly, recent research has revealed that plants, like animals, use neurotransmitters, most notably glutamate, for intercellular communication. Certain plant cells even have glutamate receptors, just like human and animal neurons do (Davis 1998). Glutamate signaling seems to play a crucial role in regulating the production of chlorophyll (Regan 149). This discovery sheds new light on plants' manufacture of psychoactive chemicals (which, as I have already mentioned, are usually close analogues of animal neurotransmitters). It was long believed that these substances played no role in plants' own metabolisms. It was presumed, rather, that they had evolved because of their effects upon animals, allowing plants to attract or repel them. But if plants also use neurochemicals, then psychoactive substances may well work directly to influence and modulate what can only be called plants' own behavior. Not only do plants exhibit tropisms toward the sun or sources of water; they also seem actively to "choose," for instance, which fungi in their roots to form a partnership with (Ehrenberg and Ross-Flanigan).[84] We can imagine, therefore, something like a plant intelligence, slow-moving by animal standards, but vast and deep. It may not be conscious as we understand the term, but perhaps it thinks and communicates in other ways. McKenna claims that plant psychedelics are an interface for such communication. Scientists are now exploring other interfaces, as well; one group, for instance, seeks "to create plants containing tiny biochip control devices in their cells capable of receiving and transmitting signals . . . Eventually, the researchers hope to figure out how to grow organic sensors that are part of the plant itself" (Kharif). For McKenna (1991), contact with plants is vitally important, for they offer us an experience of "inwardness" that we couldn't find in any other way: "inwardness is the characteristic feature of the vegetable rather than the animal approach to existence. The animals move, migrate, and swarm, while plants hold fast. Plants live in a dimension characterized by the solid state, the fixed, and the enduring." Here McKenna is guilty of idealizing the plants' slowness as a kind of stasis; actually, of course, they are always moving and metamorphosing, only at a speed that falls below our human threshold of perception. But he is right to propose vegetal "inwardness" as an antidote to mammalian hyperactivity. Think of Ryan Drum's plan for treating heroin addicts by injecting algae cells into their skulls, thus allowing them to photosynthesize their own food and drugs. The result would be *Homo photosyntheticus*, a new species of "societally harmless anthropoids nutrifying themselves in languor at a delicious if safe remove from the normal frenetic hustle of urban animal life."[85] This is a logical extension of the space of flows, which is not just a space of high speed, but one in which multiple rhythms and velocities coexist, extending well beyond the capacities of the merely human.

In the Company of Mushrooms. The biggest problem with McKenna's theories is his failure to distinguish between plants and fungi. Most biologists classify fungi as a separate kingdom, on an equal footing with animals and plants.[86] Unlike plants, fungi do not produce their own nutrients through photosynthesis, and unlike animals, they do not ingest their food. "Instead, fungi feed by *absorption* of nutrients from the environment around them" (Waggoner and Speer). Also, fungi can reproduce both sexually and asexually. In contrast to animals and plants alike, their sexual reproduction involves multiple "mating types," rather than male and female sexes. For all these reasons, McKenna's ideas about mushrooms need to be considered separately from his speculations on the vegetable mind. McKenna argues that *Homo sapiens* and *Stropharia cubensis* (the psilocybin mushroom) are coevolving symbiotes. Human evolution was catalyzed by consumption of the mushrooms; it was under fungal influence that our ancestors developed "language, altruism, planning, moral values, esthetics, music and so forth—everything associated with humanness" (Farber). Again, "all of the mental functions that we associate with humanness, including recall, projective imagination, language, naming, magical speech, dance, and a sense of religion," result from the mushroom's hallucinogenic tutelage (McKenna 1992). In return for fostering mankind's evolutionary leap, "the mushroom was able to inculcate itself into the human family, so that where human genes went these other genes would be carried" (McKenna 1983). This symbiotic synergy is so powerful, because the mushroom is already preadapted to the brains, not only of human beings, but of any "civilized forms of higher animals." The mushroom is intrinsically networked, or connected. Its "mycelial body" is "a fine network of fibers growing through the soil. These networks may cover acres and may have far more connections than the number in a human brain" (McKenna, in Oss and Oeric). Fungal mycelia, animal neurons, and Internet cables can join together to create highly distributed and massively parallel systems of information retrieval, computation, and communication. The psilocybin mushroom is therefore an apt tutelary organism for the network society, even if we are not willing to accept McKenna's wilder claims about its supposed extraterrestrial origins. In any case, we shouldn't be surprised that scientists are already trying to string together neural networks made of fungal mycelia, in order thereby to develop a "mycocomputer or Fungal Artificial Intelligence (FAI)" (Ivanov).[87]

Just-So Stories. McKenna is the first to admit that his speculations about human-fungal symbiosis are not supported by much hard empirical evidence: "I call myself an explorer rather than a scientist, because the area that I'm looking at contains insufficient data to support even the dream of being a science" (1983). But of course, there is scarcely any empirical evidence about how human language and consciousness evolved. McKenna's delirious account of human origins is of much the same order as the sociobiological explanations that Stephen Jay Gould disparages as "just-so stories," but that Daniel Dennett (212–20 and 229–51) and Steven Pinker (1997, 21–44) justify as "reverse engineering."[88] Gould complains that there is no objective way to evaluate sociobiologists' attempts to "explain form and behavior by reconstructing history and assessing current utility." Even if a story about a given adaptation is plausible and consistent, this does not guarantee that it is actually true, since "there are a multitude of potential selective explanations for each feature." Dennett and Pinker do not answer this objection so much as they dismiss it out of hand.[89] Pinker repeats again and again that "evolutionary thinking" and "the computational theory of mind" are "indispensable," and that without them "it is impossible to make sense of the evolution of the mind" (23–27). There are at least two dubious assertions here. First, Pinker implicitly claims that "evolutionary thinking" and cognitive science are the *only* theories that can make sense of the mind. They are not just the best theories we have now, but the only good theories we could *ever* possibly have. Unintelligibility is the sole alternative. Such absolutism is far from scientific. Second, Pinker assumes that, if we "make sense of" how the mind evolved, perhaps by telling a convincing story about it, then we will automatically know "how the mind works" in the present. But this is a simple logical error: the confusion of origin and purpose.

As Darwin's near contemporary Nietzsche wrote, "the cause of the origin of a thing and its eventual utility, its actual employment and place in a system of purposes, lie worlds apart; whatever exists, having somehow come into being, is again and again reinterpreted to new ends, taken over, transformed, and redirected" (1969, 76–77). Overall, Pinker's rhetoric amounts to asserting that *any* minimally coherent explanation is better than none at all. Just as, in Nietzsche's troubling words, the human will would "rather will *nothingness* than *not* will" (97), so the scientific will according to Pinker would rather invent an arbitrary explanation than not have any explanation at all. If evolutionary psychology can come up with a plausible and consistent story about how any feature of human behavior got that way, and Gould (by his own admission) cannot, then this in itself is all the validation that Pinker needs. As long as the story doesn't actively contradict the known data, it may be regarded as true. So cavalier an approach to matters of evidence and proof is a typically *postmodern* one. This is quite ironic, given Pinker's rants against "postmodernism, poststructuralism, and deconstructionism, according to which objectivity is impossible, meaning is self-contradictory, and reality is socially constructed" (57).[90] Unwittingly, and in spite of himself, Pinker is forced to the same conclusion as Borges, Derrida, Rorty, or for that matter McKenna: that our "just-so stories," or multiple, contending, a posteriori narratives, are the only "truth" we have. In contrast, Gould comes across as an old-fashioned stick-in-the-mud, who still maintains his faith in the Enlightenment ideal of objective, empirical verification.

Reverse Engineering. Pinker describes his approach as follows: "Reverse-engineering is what the boffins at Sony do when a new product is announced by Panasonic, or vice versa. They buy one, bring it back to the lab, take a screwdriver to it, and try to figure out what all the parts are for and how they combine to make the device work" (1997, 21). The stock objection to such an approach is that living organisms were not designed by engineers with foresight and goals, in the way that the Sony Walkman was. Natural selection operates without agency; it cannot look ahead, and it does not have any purpose. Pinker's equally stock response is that the sheer fact of reproductive success can stand in for purpose. Natural selection is a process of continual optimization. Even though no actual agency is involved, optimal results can be regarded *as if they were the purpose of the process.* In Pinker's words, "Darwin, and then Dawkins, made it scientifically respectable to talk about genes as having goals, because natural selection makes them act as if they do" (Pinker and Wright 2000). Respectability is one thing, however, and accuracy is quite another. Darwin himself is acutely aware both of the problems that arise when metaphors are introduced into scientific discourse, and of the impossibility of doing without them. He defends his use of the term *natural selection,* even though there is no agency that could do the selecting: "in the literal sense of the word, no doubt, natural selection is a false term," but "every one knows what is meant and is implied by such metaphorical expressions; and they are almost necessary for brevity" (1998, 109). This circumspection is exemplary, but not all of Darwin's followers have been as scrupulous in their use of metaphor. Dawkins's invocation of *selfish genes* is nearly as careful as Darwin's use of *selection.* But the same cannot be said for Pinker's assertion that "the ultimate goal that the mind was designed to attain is maximizing the number of copies of the genes that created it" (1997, 43). Here Pinker takes the metaphor of a *goal,* which works well in limited circumstances, and applies it with indiscriminate generality. The rhetoric of purpose is indeed a useful heuristic tool if we are trying to understand how the eye works, for instance, by assuming that it evolved as it did in order to enable a certain sort of vision. But Pinker's statement about the "ultimate goal" for which the mind was designed is useless. Taken in its broadest sense, it is an empty tautology: any feature of human beings that evolved through natural selection—including the "feature" that we are beings whose behavior is often shaped by cultural forces, rather than genetic ones—must, *by definition*, have served the goal of "maximizing the number of copies of the genes that created it." Taken in any narrower sense, however, Pinker's assertion is nonsense: the human mind does not serve the purpose of genetic replication in any way that is meaningfully analogous to how the human eye serves the purpose of vision.[91] Pinker can only make his argument by covertly (and perhaps unwittingly) appealing to a finality that extends beyond the heuristic use of specific assumptions of purpose. A few more steps along this road, and we arrive at the absurdities of Robert Wright, who argues in his book *Nonzero* that the purposeful force of evolution is such that "directionality is built into life... life naturally moves toward a particular end" (4). What's more, this "particular end" miraculously turns out to be the very network society that we live in today: "Globalization, it seems to me, has been in the cards not just since the invention of the telegraph or the steamship, or even the written word or the wheel, but since the invention of life" (7). The theory of evolution through natural selection abolishes purpose and agency; this is what made it so scandalous when Darwin first proposed it. But contemporary neo-Darwinism seems strangely intent on sneaking purpose and agency back in again, thus reviving the Natural Theology that Darwin overthrew.[92]

Purposiveness without a Purpose. "We invented the concept 'purpose': in reality, purpose is *lacking*." Nietzsche's maxim (1968a, 54) aggressively restates something that is already at least implicit in Kant's distinction between regulative and constitutive ideas. Kant says that "the transcendental ideas are never of *constitutive* use...On the other hand, the transcendental ideas have a superb and indispensably necessary *regulative* use: viz., to direct the understanding to a certain goal by reference to which the directional lines of all the understanding's concepts converge in one point" (1996, 619). That is to say, we can and should use the idea of a purpose regulatively, as a guide to help us make sense of how an organism functions and how its features might have evolved. It is scarcely possible for us to understand organisms in any other way. But we should always keep in mind that such a purpose is not constitutive: it is not the actual inner principle of the organism's functioning, nor is it actually present in the history of how the organism evolved. Purpose is a frame of reference that goes beyond empirical cause and effect; we do not discover it within phenomena so much as we attribute it to phenomena. This is what Kant means when he calls it a transcendental idea. Indeed, *purpose* is not just one example of such an idea. Rather, it is the privileged form of transcendental ideas in general, for using ideas regulatively means using them as heuristic tools "to direct the understanding to a certain goal." Thus, the idea of purpose is a kind of necessary fiction. It is necessary, because it alone allows us to organize vast quantities of empirical data. But it is a fiction, because it is not immanent to the field of data thus organized. In Nietzsche's terms, purpose "has only been *lyingly added*" (1968a, 36) to phenomena; it is an active interpretation, or an imposition of the will to power. But Nietzsche, no less than Kant, sees this imposition as a necessary one; we cannot imagine things being otherwise. The fictive idea of purpose provides a crucial link between Kant's epistemology (*The Critique of Pure Reason*) and his aesthetics (*The*

Critique of Judgment). Science and art alike require a certain
suspension of disbelief. We are guided in making sense of
phenomena by regarding them as if they were purposive,
even though we know that actually they are not. Similarly,
we enjoy the beauty of an object when we are able to perceive in it "a purposiveness as to form . . . without basing it
on a purpose" in actuality. Thus the objectivity of scientific
knowledge, and the disinterestedness of aesthetic enjoyment, are both grounded in what Kant calls "purposiveness without a purpose" (1987, 64–65).

Feedback. Pinker's functionalist idea of finality might be
characterized, in opposition to Kant's aesthetic one, as a
state of purpose without purposiveness. I am tempted to
call it the *anesthetic condition.* One reason that Kant sees
purpose as a regulative fiction, and not an empirical fact, is
that any explanation of a thing in terms of the purpose it
serves leads to a disturbing reversal of causality. If we say
that a thing exists *because* of the effect that it has, then "the
presentation of the effect is the basis that determines the effect's cause and precedes it" (1987, 65). By "presentation of
the effect," Kant means a conscious act of will, as when an
architect envisions the structure of a house before starting
to build it. But in the absence of actual planning and foresight, we can only speak, by analogy, of a general idea of
purposiveness: "we do call objects, states of mind, or acts
purposive even if their possibility does not necessarily presuppose the presentation of a purpose; we do this merely
because we can explain and grasp them only if we assume
that they are based on a causality [that operates] according
to purposes." Pinker, however, wants to claim more for the
idea of purpose than this. He is not bothered by the strange
inverted logic of a "theory in which *how well something works*
plays a causal role in *how it came to be*" (1997, 162). Indeed,
he finds an adaptive purpose in nearly everything, all the

more so in that there is no purposiveness—no will, no foresight, and no planning—behind it. Pinker explains the possibility of such purpose without purposiveness in terms of feedback: "one attributes a 'goal' to an entity only if it has a feedback mechanism that makes the entity approach the goal despite obstacles or perturbations" (Pinker and Wright). Pinker is describing a thermostat, which continually adjusts its heating and cooling mechanisms in order to maintain a fixed temperature. He goes on to assert that, in the same way, "natural selection is a feedback process with a kind of 'goal,' and so is human striving." It's rather odd for Pinker to describe human goals by comparing them to a thermostat, which itself can only be said to have goals metaphorically, by comparison with human intentionality. But Pinker's definition of *goal* is even more problematic. For one thing, Pinker makes the hidden assumption that the goal is always a condition of homeostatic equilibrium, rather than (for example) one of growth, transformation, or expenditure.[93] For another, his formulation begs the question of who set the thermostat in the first place, and why it was set at one temperature rather than another. From this point of view, the thermostat analogy is precisely the sort of thing that a creationist would use to argue against natural selection, and in favor of "intelligent design." Pinker evidently recognizes this, because later in the same discussion he reverses himself and writes that "natural selection itself, being a product not of a teleological process but of the physics and mathematics of replicating systems, has no right to have a goal in the way that genes or people or thermostats do." But this only adds to the confusion. Pinker now asserts, rightly, that natural selection is not like a thermostat after all, and does not have a goal. But the reason he gives for this lack of purpose is that natural selection is not itself the product of a process that has a purpose. If this argument were correct, then nothing produced by the

nonteleological "replicating system" that is natural selection could have a purpose either. As for "human striving,"
Pinker seems to place it, together with genes and thermostats, in the category of things that have goals that are assigned to them from outside. But he also asserts that "the
only processes we know of that can imbue things with
goals are human intentionality and natural selection." The
thermostat has a goal because human beings have assigned it one; presumably human beings have goals because our genes have assigned them to us, and the genes
have goals because natural selection has assigned these
goals to them. And this, in turn, seems to imply that
"human intentionality" is so by analogy to something that
is itself nonintentional. I could go on with this tedious deconstruction of Pinker's rhetoric, but I think I have done
enough to make my point. Which is, that Pinker's arguments about goal and purpose are both circular and inconsistent. They fall into the paralogisms and antinomies that
must result from any attempt to use the idea of purpose con-
stitutively instead of regulatively.

Efficiency. The theories of evolutionary biology and
free-market economics are similar in many ways. Darwin's
natural selection and Adam Smith's invisible hand are both
distributed processes that work from the bottom up, blindly,
without supervision or planning. Both generate order, or
"observed regularities" (Friedman 13), out of chaos. Both
are supposed to produce optimal outcomes, starting from
initial conditions of scarcity. Individual genes compete to
maximize their own replication, just as individual human
beings compete to maximize their utility. The aggregate
outcome of all these struggles is an overall state of equilibrium. Natural selection leads to *evolutionarily stable strategies:*
common patterns of behavior that, "once evolved, cannot

be bettered by any deviant individual" (Dawkins 1989, 69).[94] The invisible hand of the marketplace leads to *Pareto efficiency:* a condition in which no one can be made better off without someone else being made worse off. But these two theories are more than just structurally equivalent. There is also a deep affinity between them, in terms of style and affect. Practitioners of both tend to be obsessed with questions of efficiency and utility. They reject the idea that anything found in nature or society could ever be arbitrary, meaningless, or irrational. They take pride in tracking down even the slightest instances of waste, randomness, or dysfunction, and proving that the phenomena in question are actually functional, useful, efficient, and rational after all. Above all, they are always congratulating themselves for coming up with clever arguments that are as outrageous and counterintuitive as possible. Steven E. Landsburg, for instance, delights in explaining that seat belts decrease automotive safety (3–6) and that recycling paper leads to the loss of trees (81). David Friedman demonstrates that you are made better off when housing prices go down after you have bought a house (34–36). Similarly, in the field of evolutionary psychology, Randy Thornhill and Craig T. Palmer argue that rape is an efficient, adaptive strategy for men to increase their reproductive success. The only thing that makes any of these theorists blanch is the prospect of waste or of gratuitous action on the part of some "deviant individual." Everything is permitted, except for a failure of rationality. There's nothing scandalous about rape per se, but it is indeed a scandal if the rapist wears a condom, or ejaculates prematurely, or otherwise fails to fertilize his victim.[95]

Money for Nothing. Extravagance is always the problem, for evolutionary biologists and free-market economists alike. The wanton, conspicuous squandering of resources should not even be possible, according to the logic of both theories, yet we see it happening all the time. Steven E. Landsburg is appalled by the "ridiculously long tails" of male birds of paradise, "far too long for any practical purpose." He seems to believe that such extravagantly wasteful exhibitions are not found in human society, unless the government interferes with the functioning of the free market. Struck by this contrast between nature and culture, Landsburg says that "nothing in evolutionary theory either promises or delivers the spectacular efficiency of the competitive marketplace" (73). The crucial difference is that the marketplace (but not the jungle) is regulated by the discipline of prices. In the marketplace, unlike in nature, there is no free lunch; making it possible to demand payment for everything is the great triumph of human civilization. As Arnold's father puts it, in Jane Bowles's novel *Two Serious Ladies:* "What separates a man from a wolf if it is not that a man wants to make a profit?" (21). In the same spirit, Landsburg argues that "without prices, there is no reason to expect efficient outcomes. I see no analogue of prices in the origin of species, and conclude that evolutionary biology bears only the most superficial resemblance to the economics of the marketplace" (77). In fact, biologists explain the long tail on the grounds of sexual selection, rather than straightforward natural selection. Female birds of paradise prefer to mate with longer-tailed males. Males therefore compete to grow longer tails, even though this is to the detriment of their ability to fly, to feed, or to escape from predators. Indeed, it is precisely because the male's long tail is a self-crippling extravagance that it works as a signal of desirability as a mate. Such is the logic of Amotz and Avishag Zahavi's *Handicap Principle.* The male demonstrates to the female that it is so strong and healthy that it can still thrive, even with the "handicap" of a large tail. Waste and inefficiency turn out to have a second-order utility after all, insofar as they work as reliable signals. "The cost—the 'waste'—is the very element that makes the showing off reliable," and without any chance of deception (40). Just as only the richest people can afford the extravagance of large diamonds or a Mercedes, so only the healthiest birds can afford the extravagance of a large tail. The Zahavis thus present the handicap principle as a rational way "to explain ostentatious waste in nature" (39).

Ostentatious Waste. Extravagance is only a problem in conditions of scarcity. If organisms are competing for limited resources, then waste is not just something that needs to be explained; it is something that positively must be explained away. But under conditions of abundance, waste and extravagance are only to be expected. Such is the point of Nietzsche's critique of Darwin: "the general aspect of life is *not* hunger and distress, but rather wealth, luxury, even absurd prodigality; where there is a struggle it is a struggle for *power*," and not just for mere survival (1968a, 75). Or again, "in nature it is not conditions of distress that are *dominant* but overflow and squandering, even to the point of absurdity. The struggle for existence is only an *exception*, a temporary restriction of the will to life" (1974, 291–92). In short, Nietzsche says, "one should not mistake Malthus for nature" (1968a, 75). Now, Nietzsche is talking about natural history here, but his point is equally valid for political economy, for the Malthusian assumption of fundamental scarcity is as important to free-market economics as it is to evolutionary biology. Without scarcity, there can be no pressure of competition, and the quest for efficiency becomes absurd. Indeed, the capitalist obsession with scarcity—and hence, too, with efficiency, and with the struggle for survival—dates back, well before Malthus, and even before Adam Smith, to the Protestant ethic that Max Weber saw as the basis of capitalism. Today, it might seem as if this has changed. In our postmodern, globally networked world, the motor of capitalist accumulation is not carefully controlled reinvestment, but manic consumer spending, and even more manic financial speculation. It's hard for us not to laugh at something as quaint as Benjamin Franklin's maxims of frugality, analyzed by Weber as quintessential expressions of the spirit of capitalism (48ff). We are more accustomed to George Gilder's paeans to entrepreneurialism, with their celebrations of capitalist expenditure and abundance. At the same time, capitalism is most commonly reproached, by moralists on both the left and the right, for being wasteful and for stimulating useless cravings. But isn't the real problem with capitalism rather that it is never truly extravagant, never wasteful or useless enough? Everything still comes down to the bottom line: the need for profit, and immediate, short-term profit at that. The ruthless cost-cutting and downsizing measures practiced by so many corporations over the last few decades indicate that scarcity is still the governing assumption of capitalist practice as well as theory.

Unproductive Expenditure. Following in Nietzsche's footsteps, Bataille argues that Darwinian competition and capitalist scarcity are only the secondary countereffects of a more general and more primordial "superabundance of biochemical energy and growth" (1988, 27). In his early essay, "The Notion of Expenditure" (1933, translated in Bataille 1985), Bataille insists that "human activity is not entirely reducible to processes of production and conservation." Such activities as "luxury, mourning, war, cults, the construction of sumptuary monuments, games, spectacles, arts, perverse sexual activity" serve no useful, economic purpose; rather, "the accent is placed on a *loss* that must be as great as possible in order for that activity to take on its true meaning" (118). Economic rationality cannot account for these practices. Explanations in terms of scarcity and efficiency are always inadequate, for there is something in human life and society that pushes beyond the limits of "the principle of classical utility" (116). In a certain sense, however, Bataille himself is still writing from the point of view of such a principle: he only defines expenditure negatively, as an exception to the law of value, and as something that escapes utility. Unproductive expenditure, like beauty according to Kant, is fundamentally *disinterested*. But it is only in his later book, *The Accursed Share* (1949, translated as Bataille 1988), that Bataille is able to give a positive account of expenditure. In this text, he traces it all back to the infinite prodigality of the sun: "Solar energy is the source of life's exuberant development. The origin and essence of our wealth are given in the radiation of the sun, which dispenses energy—wealth—without any return" (28). The problem that living organisms face is how to absorb this superabundance of energy. Much of it is used for accumulation and growth. But there is always more energy streaming out from the sun than can be profitably employed in this way. Eventually, the earth is entirely saturated; "it is the size of the terrestrial space that limits overall growth" (29). Energy that cannot be absorbed any longer has to be wasted or squandered somehow. If I cannot get rid of it, I will explode—which is itself an extravagant form of squandering. "The dominant event" in the history of life on Earth "is the development of luxury," and particularly of "the three luxuries of nature: eating, death, and sexual reproduction," which all waste energy in spectacular ways (33). Darwinian (and capitalist) competition is not driven by any scarcity of resources, but rather by an *excess of potential* on the part of living organisms (and enterprises): they are never able to find sufficient opportunities to expend all of their potential. Neither in the biosphere nor in the human economic sphere is scarcity given in advance. Rather, scarcity is something that always needs to be produced and imposed. What a particular individual or group experiences as a "deficiency of resources" is just the secondary effect of an overall "excess of resources," when regarded from a more general point of view (39). Or to put the same point in another way: scarcity is only a problem for closed systems, which try to maintain a state of homeostatic equilibrium by sealing themselves off from the outside world. But when it comes to open systems—or what Ilya Prigogine and Isabelle Stengers call *dissipative structures*, systems that exist dynamically, in far-from-equilibrium conditions, and that are traversed by vast flows, exchanges, and expenditures of energy (12–14)—the experience is rather one of surplus and exuberance.

Potlatch. Bataille argues that capitalism differs radically from previous forms of human society in the way that it tries to manage and limit expenditure. Precapitalist cultures have gift-based economies, organized around rituals like the *potlatch* of the Northwest Coast Indians, in which massive stores of wealth are given away, or even destroyed (1988, 63–77). Archaic gift exchange has little in common with the proto-capitalist practices of barter and mercantile trading, but it is equally far from the "hippie dream" of freely exchanged information embraced by Internet pioneers, open-source software enthusiasts, and copyright violators, and so savagely derided by K. W. Jeter. Following Marcel Mauss's famous study *The Gift* (1967), Bataille analyzes the ambiguity at the heart of practices like potlatch. On one hand, such ceremonies involve ostentatious expenditure. They are devoted to "the dissipation of useful wealth," rather than to its acquisition (Mauss 1967, 68). In this manner, they affirm life's exuberance, celebrating the extravagance of loss. But on the other hand, these ceremonies are anything but gratuitous, in terms of their purposes and outcomes. They are vehicles both for creating alliances and for waging war symbolically. The potlatch ceremony is the arena for the most intense power struggles and the most vehement rivalries. The gift-giver acquires "prestige, glory, and rank," in proportion to the extent of his self-inflicted loss (71). The recipient is burdened with an onerous debt and obliged to reciprocate even more extravagantly (70). So ostentatious and reckless a squandering of wealth is neither stable nor efficient. It cannot be explained in terms of what we think of as economic rationality. It involves, Mauss says, "a state of perpetual economic effervescence" (70) and a mechanism of exchange that has nothing to do with any pricing system. A gift is returned with interest, for instance, "in order to humiliate the original donor or exchange partner and not merely to recompense him for the loss that the lapse of time causes him" (73). Thus archaic economics is ruled by "a notion neither of purely free and gratuitous prestations, nor of purely interested and utilitarian production and exchange; it is a kind of hybrid" (70). This is why Bataille finds it paradoxical. Archaic exchange is based in the act of negating economic utility. But this very negation turns out to be useful nevertheless, in a noneconomic, higher-order way. The splendor and misery of potlatch is therefore something like a human analogue to the animal behavior explained by the Zahavis' handicap principle. In society as in nature, we can account for ostentatious waste in terms of the prestige that it confers. But there is one critical difference between the handicap principle and Bataille's General Economy. For the Zahavis, Malthusian scarcity remains the bottom line. Abundance is only relative; the animal that has more energy to spare than its rivals is able to use this relative advantage as a weapon in the struggle over scarce resources. For Bataille, however, abundance always comes first. Potlatch presupposes a surplus, rather than a state of scarcity. It is not a roundabout way of fulfilling basic needs, and still less of promoting reproductive success, but a direct way of dispersing and finally destroying "objects of luxury" (76). The giver's prestige is only a temporary achievement; it is soon lost in the greater movement of exuberant dissipation. Expenditure can always be accounted for in terms of utility, but there is always something more to it, as well.

As Shameful as Belching. Bataille laments that, in the global capitalist economy, "the great and free forms of unproductive social expenditure have disappeared . . . Everything that was generous, orgiastic, and excessive has disappeared; the themes of rivalry upon which individual activity still depends develop in obscurity, and are as shameful as belching" (1985, 124). In this way, Bataille adds a new dimension to the familiar Marxist account of how capitalism appropriates, privatizes, and reifies whatever it inherits from previous social forms: not just material objects and the means of production, but all human relations as well. Where the Tlingit and Haida had potlatch, the Aztecs human sacrifices (Bataille 1988, 45–61), and the ancient Romans Saturnalia and gladiatorial contests, we must make do with pornographic Web sites, or strip clubs with individual viewing booths. Even our vicarious enjoyment of spectacle is privatized, rationalized, and limited in scope. The closest we come to a public, collective display of expenditure today is the Super Bowl, World Wrestling Entertainment matches, and media coverage of catastrophes (I write these lines in the aftermath of the World Trade Center bombings of September 11, 2001). The great paradox of capitalism is the reverse of the one that haunts archaic economies. What needs to be explained is no longer the fact that spontaneous acts of expenditure, like sacrifices and gifts, also turn out to serve useful purposes for the people who perform them, but rather the fact that an economy entirely given over to

utilitarian calculation, or to "rational choice," still continues to express—if only underhandedly, and by inversion—
the delirious logic of unproductive expenditure. Indeed,
the free market system seems grounded in denial. The more
capitalism revels in abundance, the more it demands efficiency. The more it lowers costs of production, the more it
imposes the harsh discipline of prices. The more choices
and options it offers us, the more it compels us to make
definitive and irrevocable decisions: at every turn, the demand for an exclusive *either/or* replaces the coziness and
ease of *both/and*. In short, even as it produces greater material wealth than ever before in human history, capitalism
also continually manufactures scarcity and want.

Explosion. Classical economics claims, of course, that
efficiency and market discipline, in the face of scarcity, are
what give birth to the surplus and thereby permit the expansion of overall social wealth. But Bataille's General Economy argues precisely the reverse. Abundance is not the result of capitalist investment, but its impetus and cause. It is
what makes capitalist economic relations possible in the
first place. The great historical novelty of capitalism, and
probably the secret of its success, is that it was the first economic system to dissimulate the process of expenditure
that fuels it and turn that process back upon itself. Where
other economic systems tend to discharge their surpluses
directly and spectacularly, capitalism repeatedly tries to
defer any such display of expenditure. Instead, it reinvests
as much of the surplus as possible in its own reproduction
on an expanded scale. It develops new technologies to stim-
ulate growth. "Human activity transforming the world
augments the mass of living matter with supplementary
apparatuses," and this process in turn "considerably increases the resources of available energy" (Bataille 1988,
36). The material basis of life is intensified and expanded,

particularly through what Marx calls the extraction of relative surplus value (Marx 1992, 429–38 and 643–54). The earth now supports a greater density of human life and accumulated wealth than it ever did before. But this strategy of intensification only works up to a certain point. Its very success leads it to the brink. New technologies "always have a double effect: Initially, they use a portion of the surplus energy, but then they produce a larger and larger surplus. This surplus eventually contributes to making growth more difficult, for growth no longer suffices to use it up" (Bataille 1988, 36–37). The point of total saturation is eventually reached. Capitalism is strangled by its own luxuriant growth. It is no longer able to consume its own excess. Traditional Marxist economic theory calls this a crisis of underconsumption, or overproduction. Workers may double as consumers, but in the latter role they cannot afford to purchase all the goods that in the former role they have produced. The energy that cannot be absorbed in reinvestment or growth has to be discharged in some other manner. The system is driven, in spite of itself, to the convulsions of unproductive expenditure. That is why capitalism is explosive. It cannot maintain a steady state of prosperity, but always needs to push its boundaries further outward. As Deleuze and Guattari put it, "the deepest law of capitalism" is that "it continually sets and then repels its own limits, but in so doing gives rise to numerous flows in all directions that escape its axiomatic" (1987, 472). Crisis is therefore essential to the capitalist enterprise. Marxists have hoped for well over a century to see capitalism collapse from its self-generated crises, which is to say from its internal contradictions. But the trouble is that these very crises and contradictions are *also* what keep capitalism alive. A crisis is the only thing that can rejuvenate the system when its inability to absorb its own energy has led it into an impasse. That is why capitalism is, strictly speaking, *undecid-*

able (471–73). Its crises are always generating "new flows in all directions": these flows offer us new possibilities for liberation, but they also open up new territories for capital to conquer. It is hard to see any end to this process. When the surface of the earth has finally been exhausted, and no further intensification is possible, capitalism will then have to expand into outer space—as Ken MacLeod suggests (2000a, 146–47).

A Postmodern Metaphysics. Crisis, undecidability, expenditure: these are the metaphysical concepts proper to networks and the network society. Modernity sought to eradicate all metaphysics, but it discovered, in spite of itself, that this is impossible—or better, that it is an endless task. Neither positivism, nor linguistic analysis, nor mathematical logic, nor Heidegger's interrogation of Being was able to exorcize the ghost of transcendence. Twentieth-century thought finally rediscovered what Kant already knew: that the "illusions of reason" can be criticized and compelled to acknowledge their limits, but they cannot be altogether eliminated. This is why Foucault urges that, "instead of denouncing metaphysics as the neglect of being, we force it to speak of extrabeing" (1998, 347). Metaphysics is not about objects in the world, in their presence and positivity. It is rather about the spaces between things, their zones of fuzziness and indiscernibility. It is about what Foucault, following Kant, calls "nonpositive affirmation" or "the testing of the limit" (1998, 74). And it is about the interstices of the network. "A line of becoming," Deleuze and Guattari say, "is not defined by points that it connects, or by points that compose it; on the contrary, it passes *between* points, it comes up through the middle" (1987, 293). Similarly, the hacker heroine of Pat Cadigan's science fiction novel *Synners* has to maneuver in between the nodes of the network, in order not to be wiped out by a virus: "if you couldn't walk on the floor, you walked on the ceiling. If you couldn't walk on the ceiling, you walked on the walls, and if you couldn't walk *on* the walls, you walked *in* them . . . If you were walking in the walls, and the walls had black holes, you had to be something that a black hole wouldn't recognize as existing" (351). The Marxist idea of crisis, Bataille's notion of expenditure, and Deleuze and Guattari's concept of undecidability, are all ways of walking in between—of invoking *extrabeing*—in order to oppose the solidity, the inertia, and the seeming self-evidence of the actual. They are ways of finding ambiguous points of potential, gaps in the linear chain of causality, unexpected openings to new, emergent processes. Where deconstruction can only see undecidability as the *aporia* of rational thought, leading to a paralysis of the will, Deleuze and Guattari rather welcome it as a stimulus to both thought and action: "the undecidable is the germ and locus par excellence of revolutionary decisions" (473). The absence of any cognitive grounding for our actions is precisely their condition of possibility. In the same way, there's nothing predetermined about a crisis and no guarantee that any given crisis will lead to radical change, rather than to nothing at all, or even to more misery and oppression. But crisis remains the condition of possibility for change, the metaphysical extravagance that alone can open up chinks in the otherwise impenetrable armor of the real.

Crisis Energy. Isaac Dan der Grimnebulin, the hero of China Miéville's *Perdido Street Station,* is a scientist who works on *crisis theory.* Isaac believes that "motion is part of the fabric of ontology" (169), that "it's in the *nature of things* to enter crisis, as part of what they are . . . things are in crisis, just as part of *being*" (170). For crisis is a free-floating potential that undergirds all forms of energy. It's the primordial condition out of which everything emerges and back to which everything ultimately returns again. Isaac's crisis theory stands in opposition to the mainstream dogma that regards motion and change as *"pathological...deviations* from an essential state" of identity and rest (169). Isaac is thus on the side of Heraclitus against Parmenides, of Nietzsche against the Platonic tradition, of Marxist dialectics against positivism, and of chaos and complexity theory against atomistic reductionism. In the course of the novel, Isaac puts his metaphysics into practice by building a *crisis engine.* This is a machine for unleashing—and controlling— the power of transformation. The trick is to devise a way of "taking things to the point of *crisis*" (169) and then keeping them there. It involves a sort of mathematical calculation that is itself "profoundly unstable...paradoxical, unsustainable, the application of logic tearing itself apart" (633). But if you can reach that paralogical crisis point, then "the tapping of [an object's] crisis field actually increases its crisis state"; the result is "an endless feedback loop, which means a permanent font of energy" (238). The results of deploying the crisis engine are uncertain, but it could lead to revolutionary change. No wonder Isaac's invention badly frightens the rulers of New Crobuzon; the city's business elite is far more disturbed by the crisis engine than it is by the escape of the slake-moths (406). Any particular crisis is something that they can deal with, and perhaps even benefit from. But a generalized state of crisis is something else again; it's a point at which we are suddenly able to think (and to demand) everything, even the impossible.

Biological Exuberance. The metaphysical laws of General Economy apply, not just to human economies, but to the entire biosphere. Bataille refuses to distinguish between nature and culture or between ecosystems and economies. The same outlandish flows of energy traverse human societies and biological populations alike. The only privilege Bataille accords to humanity is a quantitative one: "man is the most suited of all living beings to consume intensely, sumptuously, the excess energy offered up by the pressure of life to conflagrations befitting the solar origins of its movement" (1988, 37). We practice unproductive expenditure, and generate waste, on a far greater scale than any other organisms; that is our chief distinction on this planet. The luxurious squandering of resources is the crowning achievement of human civilization. Of course, we usually tend to misrecognize this fact. Evolutionary biologists and psychologists ignore expenditure as egregiously as free-market economists do. The virtue of Bruce Bagemihl's pathbreaking study *Biological Exuberance* is that it brings "the extravagance of biological systems" (252) to the foreground as never before. Bagemihl exhaustively catalogs naturalists' reports of same-sex copulations, and other nonreproductive sexual behaviors, in an amazing range of mammal and bird species. Whether it is male ostriches courting and mounting other males (621–26), male orcas nuzzling each others' genitalia (349–52), or female bonobos masturbating one another to climax (269–75), animals seem to spend an extraordinary amount of time and energy on such activities. Bagemihl concludes that "homosexual" and "transgendered" behavior is by no means exclusively a human phenomenon. Animal behavior of course includes, but also extends far beyond, the mere necessities of reproduction: "the equation of life turns on both prodigious fecundity and fruitless prodigality" (255). It may be that much of animals' nonreproductive sexual behavior in fact serves a secondary purpose, like establishing rank within the group or creating and strengthening social bonds among individuals; the logic would be much like that of potlatch. But the question is not whether evolutionary biologists are ingenious enough to devise arguments to demonstrate that all these instances of animal homosexuality are indeed adaptive and actually do correlate with increased reproductive success. For the real issue is not whether such adaptationist explanations are possible—which they are, almost by definition—but whether they are sufficient and complete. The events of natural history, like those of human history, can be narrated in many different ways. They can be explained from a perspective of scarcity and need, but also from one of "fruitless prodigality" and exuberance. I don't think that we can meaningfully choose between Malthusian/Darwinian adaptationist explanations, on the one hand, and Nietzschean explanations in terms of "overflow and squandering," on the other. Rather, we must be attentive to both of these at once. Darwinian natural selection, as Lynn Margulis and Dorion Sagan point out, is "a strictly subtractive process" (2002, 68). It marks a kind of inner limit to change: a minimum degree of constraint, a reductive pressure, without which evolutionary adaptation might not happen. But the "creative force" that drives evolution is "life's incessant tendency to have its own way, to create and expand," which Margulis and Sagan identify with Nietzsche's will-to-power (67). This marks a kind of outer limit of change: a maximum degree of potentiality, a plenitude of bodies and behaviors in continual ferment and mutation, providing the diversity that is a necessary condition if evolution is to happen at all.

I Want to Believe. The first question for evolutionary theory is that of the *mechanism of competition*. Is selection driven by Malthusian/Darwinian scarcity, or by Nietzschean/Bataillean prodigality and exuberance? But there is a second question as well, whose answer does not necessarily correlate with that of the first. This is the question of the *units of selection*. Does evolutionary competition take place among genes, or among memes? Memetic or cultural selection need not have any underlying genetic basis. As Richard Dawkins himself puts it, "a meme has its own opportunities for replication, and its own phenotypic effects, and there is no reason why success in a meme should have any connection whatever with genetic success" (1982, 110). The distinction is crucial: Dawkins is a Darwinian reductionist, but he is not a genetic reductionist. He is not interested in asserting the primacy of the biological realm, but only in defining the *logic of selection*, wherever it may manifest itself. This comes out most clearly, perhaps, in Dawkins's view of religion. What is it that makes us "want to believe," like Fox Mulder in *The X-Files?* Orthodox sociobiologists tend to argue that the propensity for religious belief is somehow hardwired into our genes. It must have some adaptive value for humanity—or at least, it must have had such value in the past. Thus Edward O. Wilson surmises—without any evidence aside from the widespread existence of religious belief in human societies—that "acceptance of the supernatural conveyed a great advantage throughout prehistory, when the brain was evolving" (262). Steven Pinker is unwilling to go quite this far, but he is also unwilling to drop the dogma of genetic primacy. He suggests that, while religious belief may not be advantageous in itself, it must be an inevitable by-product of genetically driven processes that are. Religious belief arises spontaneously, he speculates, "from a system of [mental] modules designed to figure out how the world works" (1997, 556). Dawkins (1991), however, will have none of this. He argues that "the mental virus of faith" has no adaptive value for us whatsoever. Just like biological and computer viruses, it is a parasite that promotes its own replication at our expense. Belief is only advantageous for the religious meme itself, not for the human bodies and brains that are its host. Far from being a consequence of our drive to understand the world, the virus of faith systematically cripples that drive to preserve itself from skeptical questioning. Dawkins gleefully cites Tertullian's notorious maxim, *certum est quia impossibile* ("it is certain because it is impossible") and recalls Lewis Carroll's White Queen, who boasts of having made herself believe "as many as six impossible things before breakfast." The positivistic Dawkins exhorts us to reject such impossibilities. Fox Mulder, in contrast, urges us always to remain open to "extreme possibilities": a Kantian regulative maxim that guides his work on the X-files. But Dawkins and Mulder alike understand that the issue that divides them—that of the potentialities generated by selection—is both logically and pragmatically independent of the question of nature versus culture.

Nature and Culture. However, memes (assuming they exist) are unlike genes—and therefore culture is unlike nature—in at least one crucial respect. Memes and their hosts (human minds) have nothing that corresponds to the distinction between the germ plasm (the actual sperm or egg cells, through which alone inheritance takes place) and the soma (the rest of the body) in biological organisms. This difference has several important consequences. First, Lamarckian inheritance—the passing on of acquired traits—is the rule for memes, whereas it is impossible for genes. Second, group selection is a much more plausible mechanism for memes than it is for genes (Blackmore 197ff). Third, memetic variations are not culled as ruthlessly by the forces of natural selection as are genetic differences. And finally, memes do not spread by descent, as genes do; rather, memes are propagated laterally by a sort of contagion, or better, by a process of virulent imitation.[96] The result of all these differences is that memes vary much more widely, and change much more rapidly, than genes do. This is one reason for human sexual diversity. Our multifarious forms of nonreproductive sexual behavior comprehend and exceed those of all the other species cataloged by Bagemihl. It is as if, by simulating "biological exuberance" memetically, we can explore far larger areas of the "possibility space" or "fitness landscape" than is possible by genetic means alone. For this reason, we can say that culture inclines more to prodigality and excess, while nature inclines more to Malthusian scarcity, even if both sorts of processes are at work in both realms. Memes, far more than genes, are apt to transform themselves through the "rhizomatic" processes of "aparallel evolution" and "transversal communication" so often cited by Deleuze and Guattari (1987, 10–11).

Virulent Imitation. In her book *The Meme Machine*, which attempts to put the science of "memetics" on a solid theoretical basis, Susan Blackmore distinguishes between simple forms of "social learning" that are common among animals and the process of imitation in a strong sense (47ff). Birds and mammals often copy methods of catching prey from other individuals of their own species, for instance. But this sort of copying does not create new behaviors; it merely leads to new applications, in new situations, of already existing, innate behaviors. Only human beings have "the skill of generalized imitation," which allows them to "invent new behaviors of almost unlimited kinds and copy them on to each other." This means that "only humans are capable of extensive memetic transmission" (50–51). All forms of "social learning," animal or human, operate on a model of contagious propagation; beyond a certain threshold of density, and in the absence of countervailing factors, the learned behaviors tend to spread at an exponential rate. But human cultural transmission is far more complex than this, for it also involves explicit instruction on the one hand, and widespread rehearsal and repetition on the other. That is to say, "memetic transmission" presupposes at least two sorts of social practices: first, giving and receiving orders, and second, mimesis. The weakness of Blackmore's memetic theory is that she is unable to give an adequate account of either of these intrinsically problematic processes. Consider the act of giving orders. This is one of the basic functions of language, as Wittgenstein, Deleuze and Guattari (1987), and Morse Peckham all in their own ways argue. Who has the power to give orders? And how does one make sure that the orders one gives are understood and obeyed? How does one bridge what Wittgenstein calls the "gulf between an order and its execution," the gap between the indeterminacy of interpretation, and the finality of action? Human society is predicated upon what Peckham calls "cultural redundancy" or "the constant reiteration of instructions for behavior" (169). Only this incessant repetition makes possible the "channeling" and regularization of human activity, in spite of "the brain's capacity to produce random responses" (166). Now, natural selection can only take place when there is both a relatively high degree of copy fidelity *and* a mechanism of variation. In human culture, the brain's innate capacity for randomness provides the variation. But for this very reason, copy fidelity is difficult to obtain. Peckham observes that interpretive stability "can ultimately be maintained only by the application of force in the form of economic deprivation, imprisonment, torture, and execution" (243). As for mimesis, it is similarly double-edged. It's the human drive to imitate that allows Blackmore to posit the existence of memes, or units of cultural replication, in the first place. But widespread and virulent imitation is not just (as Blackmore seems to think) a copying mechanism. It is also (just like potlatch and other forms of gift giving) a channel for competition and contestation, giving rise to the violent "mimetic rivalry" theorized by René Girard and to the subversive, destabilizing "mimetic excess" described by Michael Taussig. Mimetic violence and mimetic contagion are active processes, even if they are not consciously chosen ones. Like power relations according to Foucault, they are "both intentional and nonsubjective" (1978, 94), which is to say (in Kantian terms) that they are purposive without having a purpose. Mimesis produces effects, and provokes interactions, that cannot be reduced to the Malthusian/Darwinian paradigm of differential selection amidst conditions of scarcity.

Funkentelechy versus the Placebo Syndrome. The most interesting thing about Blackmore's argument is that she invokes memes to explain precisely those aspects of human culture that sociobiology cannot well account for. Evolutionary psychologists have engaged in tortuous reasonings to argue for the alleged genetic advantages of practices like altruism and nonreproductive sexuality. We have seen how Dawkins rejects such arguments in the case of religion. Blackmore simply generalizes Dawkins's logic. Things like voluntary celibacy (138–39), the deliberate use of birth control (139ff), or the belief in alien abduction (176ff) cannot possibly be motivated on the grounds of promoting increased reproductive success. There has to be some other way to account for them. There are really only two choices at this point: the creative effervescence of what George Clinton calls *funkentelechy,* or the viral dread of the *placebo syndrome.* Either we throw out the notions of purpose and function altogether, as Bataille does with his idea of unproductive expenditure, and Peckham does with his observation of the human brain's capacity for randomness (162ff), or else we can try to rescue purpose and function, as Blackmore does, by saying that they do indeed exist in such cases, only not for us. For the genes' or memes' interests are not our own. These "selfish replicators" work just to perpetuate themselves, even if this be at our expense. The celibacy meme, for instance, is good at spreading itself, because it causes its host—a celibate priest, let's say—to spend all his energy proselytizing for celibacy, instead of expending that energy on producing and raising children. The celibacy meme thus has the effect of cultivating "an ongoing supply of new recruits to celibacy" (Blackmore 139). In this way, the meme is self-reinforcing (or highly redundant, in Peckham's terminology). It tends to spread among human populations, therefore, even though it contradicts our (genetic and memetic) sexual impulses, and even though the ultimate effect of its success would be to eliminate human beings altogether (and thereby, of course, abolish the meme itself as well). Such is the fatal logic of memetic proliferation.

Trust the Force. Dawkins makes a crucial distinction between *replicators* and *vehicles*. Replicators are genes or memes: "the fundamental units of natural selection, the basic things that survive or fail to survive, that form lineages of identical copies with occasional random mutations." Vehicles, in contrast, are "large communal survival machines," the phenotypic envelopes that carry those replicators (1989, 254). The vehicles for genes are individual bodies, or populations, or entire species. Analogously, the vehicles for memes might be individual minds, or interest groups or communities of various sorts, or even entire societies. The reason Dawkins makes the distinction between replicators and vehicles is to clear up an ambiguity about the question of *levels of selection*.[97] It's a real question as to whether natural selection takes place more often through competition between individuals, as Dawkins himself is inclined to believe, or rather through competition between populations or species ("group selection"), as Stephen J. Gould has frequently argued. But Dawkins's point is that, whatever the level of phenotypical competition, it is ultimately the replicators themselves, and not their vehicles, that are being differentially selected. The debate over group selection therefore has no bearing on the question of "selfish replicators." They are two entirely separate issues. And indeed, as I have already suggested, it is thanks to this distinction that we can rightly say that we are not our genes and that we are not our memes, even if we are their products and their hosts. This is the point, however, where Blackmore's argument takes a sudden strange turn. After citing Dawkins's exhortation that we "rebel against the tyranny of the selfish replicators," she suddenly asks: "But who is this 'we'?" (219). She goes on to argue that the self is nothing more than a "memeplex" (219–34), a cluster of closely associated memes. The very idea of selfhood is an illusion, and so is the idea of making free or voluntary choices. For in every case, "the 'me' that could do the choosing is itself a memetic construct: a fluid and ever-changing group of memes installed in a complicated meme machine" (241–42). Blackmore ends *The Meme Machine* by advocating a sort of *Star Wars* "Trust the Force" philosophy: "to live honestly, I must just get out of the way and allow decisions to make themselves . . . Actions happen whether or not 'I' will them . . . You do not have to try to do anything or agonize about any decision . . . The whole process seems to do itself" (244). Don't worry, be happy, Blackmore seems to be saying; the memes themselves will take care of everything. In this account, the very idea of Darwinian selection has vanished; there is neither competition among memes nor difference of interest between the memes and their vehicles. To say that everything in our mental, cultural, and social worlds is a meme is like saying that everything in our bodies is a gene. If you abandon one side of Dawkins's distinction between replicators and vehicles, then the other side also becomes meaningless. In generalizing Dawkins's idea of memes, Blackmore thus ends up inadvertently abolishing the very logic behind it. She empties out difference and conflict, erases singularity, and leaves us with the melancholy conclusion of nothing but a self-regulating and self-perpetuating network of collaborating memes.

Cyberspace versus Homunculus. The same conflict seems to repeat itself in various forms, like a tensile pattern inscribed in many different materials. There's an opposition between the network and the self, but also between proliferation through sampling and copyright control, between the mind-expanding revelations of LSD and the narrowly focused power trip of cocaine, between artificial intelligence and all-too-human emotions, between the "discipline of the marketplace" and the naked singularity of suffering flesh (Agamben), between replicators and vehicles (Dawkins), between expenditure and utility (Bataille), between Mickey-san and Ded Tek (Misha), between Eden-Olympia and the immigrant towns below it (Ballard), between the space of flows and the space of places (Castells), and between vampire-capital and zombie-consumption. In the words of Rainier Funke, it's the battle of Cyberspace versus Homunculus. On one hand, the seamlessness and "unthinkable complexity" (Gibson 1984, 51) of the system of posthuman, virtual existence; on the other hand, a stunted and shrunken residue of humanity, possibly monstrous. These two sides are never equal adversaries. Indeed, they are not even on the same level. The second term is often a subset, or a product, or an effect, of the first. For instance, the vehicle is constructed by the replicators, the self is a node in the network, and the space of places is structured by the differential technological and economic investments of the space of flows. More generally, the first term is logically on a metalevel in relation to the second. Thus expenditure and utility are not really opposed terms, because the *restricted economy* of the latter is finally just a particular instance of the *general economy* of the former (Derrida 251–77). In short, Homunculus is born out of the Matrix—which is to say the womb—of Cyberspace. The conflict, then, is not between two systems, or two sets of rules, but rather between a system and one of its own terms, or between a rule and one of its implementations. Homunculus is an *exception* to the laws of Cyberspace, but this also means that it is only *within* Cyberspace that Homunculus has its exceptional status. The logic of this conflict is a "deconstructionist" one, or what Gregory Bateson (271–78) calls a "double bind"; it violates Russell's Theory of Types by putting a statement in conflict with a metastatement that refers to it. And indeed, this paradoxical structure is what makes current struggles over globalization so notoriously difficult to parse. The struggle of Cyberspace versus Homunculus can be read, ambiguously, both as homogenizing, universalizing capitalist power versus local resistance, and as cosmopolitanism versus oppressive cultural conformity. Biotech companies versus indigenous peoples, but also the World Trade Center versus Al-Qaeda. This is the politics of the twenty-first century. As Castells (1997) suggests, the globalizing, capitalist "network society" is contested by a multitude of "local" affirmations of identity. These affirmations range from gay and feminist movements to ecological activism to various nationalisms and religious fundamentalisms. But what makes the situation even more complex is that the two sides borrow from one another. Global flows of capital invest and commodify all of these local identities, even as the identity movements themselves take the form of networks, the better to contest the networked global order.

All Politics Is Local. "The Universe is overdetermined": this is a well-known saying in the galaxy of Samuel R. Delany's science fiction novel *Stars in My Pocket Like Grains of Sand* (164). The book presents us with a cosmos of "(approx.) six thousand two hundred worlds" inhabited by human and other sentient beings (142). It's impossible for anyone to be familiar with more than a tiny fraction of all these planets and cultures. As Carl Freedman puts it in his fine reading of the novel: "size on such an order of magnitude as this text suggests must involve a literally awesome *complexity*. The attempt to think, and even more, to imagine such a huge field of difference is almost impossibly daunting and can result in a sort of mental vertigo" (2000, 150). Freedman's formulation is precise, and not merely impressionistic. *Stars in My Pocket* gives a vivid portrayal of at least a few of its worlds: the book is rich in sensory and material detail, and in what anthropologists call the "thick description" of social practices ranging from styles of dress, to the division of labor, to table manners. But the novel also makes us aware of just how much context we are missing, and how many details are absent. The narrator is always reflecting on how surprisingly diverse the worlds are, and how little she[98] actually knows. What applies in one particular place and time does not necessarily apply in another. For instance, in many cultures on many different worlds, a nod of the head means "yes," just as it does for us, but we cannot take this signification for granted, because there are also many cultures where a nod means "no," or a number of other things (149, 181). The characters of *Stars in My Pocket* spend most of their time negotiating the vast cultural and social differences that they encounter. You cannot even insult somebody properly across cultures without there being a prior transcultural understanding of what behavior constitutes an insult (312, 317–18). In a cosmos of such diversity and complexity, all signification and all understanding is irredeemably local, and every point of view is unavoidably partial. Indeed, the impossibility of transcending the local is the universal condition of Delany's cosmos, the one feature that is not itself merely local. That is to say, this impossibility is precisely what constitutes the global (the transglobal? the galactic? the universal?) in contradistinction to the local. In such a cosmos, globalization or universalization is entirely real, but it is something that cannot be grasped all at once. It can only be experienced by traversing the network, by moving continually from node to node, as each node leads to—and is overdetermined by—many other nodes.

Visible and Invisible Persons Distributed in Space. In *Stars in My Pocket Like Grains of Sand,* the vector of universalization—the process that links the disparate worlds and cultures to one another—is the continual, networked flow of information. A shadowy bureaucracy, aptly known as the Web, is "the interstellar agency in charge of the general flow of information about the universe in many places" (87).[99] Or more precisely, "the Web is information" (148): its bureaucrats and protocols and security procedures are the material form that the network—the flow of information—takes. As the Web gathers and disseminates its data, it imposes its own topography upon the cosmos, a "worldwide informational warp" (60), or perhaps an *informational curvature* roughly analogous to the gravitational curvature of space-time. Information is unevenly distributed in space; it is denser here, and rarer there. There are even places that the Web does not penetrate at all. At any particular point, the information actually available is partial and incomplete. Some information can be accessed directly via neural implants; the trouble is that "it's often ten years out of date" (243). Other information is inaccessible for the moment because it is "undergoing extensive revision" (94), or because "all information channels are currently in use and/or overloaded" (74). In still other cases, requests for certain bits of information will lead you into "a really astonishing run-around of cross-references that, as you go pinning them down, will finally result in your question being declared nonsense" (94). Even worse, ask the wrong question and "your security status automatically changes" to one of reduced access (95). Apparently, nothing is censored outright, but some information is deliberately made very difficult to obtain. These restrictions are "frowned upon by the Web but sometimes are necessary with information the kind of commodity it's become" (95). In short, the price of universalization is differential access and imposed scarcity. Information is always *about* particular localities, and even its distribution varies locally. It is only the actual *flow* of information—the process by which it is disseminated, but also homogenized and filtered—that can be described as universal, rather than local.[100]

General Information. In the universe of *Stars in My Pocket Like Grains of Sand,* the Web controls how information is collected and disseminated. It is the master of the universalizing process that Marx calls circulation in order to realize surplus value, and that Deleuze and Guattari call "the disjunctive synthesis of recording" (1983, 12). But there is always a price to be paid for any system of universal equivalences; at the very least, there are problems—or glitches (Delany 1985, 71)—of *translation*. Nothing can be rendered precisely when languages and social norms and customs differ. At one point Delany's narrator quotes the maxim, *"Poetry is what is avoided as it is surrounded by translation"* (130). And indeed, the Web's distribution, orientation, and translation service, known as General Information, often seems to be characterized as much by what it avoids as it is by what it carries over from one world to another. But even leaving these issues of translation aside, there's the further question of "what information is or is not acceptable in this or that part of this or that world" (88), that is to say, of how any amount of information, once it has been translated, recorded, and distributed, is actually used. This is the process by which the universal returns back to the local, and impersonal process is transmuted into subjective experi-

ence. Raw information is selectively made into local meaning, and particular social values and norms are thereby generated and reinforced. The Marxist tradition identifies this
process with the workings of ideology, and (in Althusser)
with the interpellation of the subject. Deleuze and Guattari
describe it as *surplus-enjoyment*, a "conjunctive synthesis"
of consumption/consummation out of which a euphoric
"residual subject" is produced (1983, 1721). But in any case,
the Web does not—and indeed cannot—directly determine
this additional process of signification and subjectification.
Other forces—forces of interpretation and selection—necessarily come into play. And these other forces are intrinsically local; they stubbornly resist the pull of any system of
universal equivalences.

Family and Sygn. *Stars in My Pocket Like Grains of Sand*
is largely concerned with the "schism" (198) or "war" (295)
between two such forces of interpretation and selection,
known as the Family and the Sygn. These forces are vari-
ously described as ideologies, cultural tendencies, and religions; they are rival paradigms of meaning, methods of gov-
ernance, and definitions of "just what exactly a woman [i.e.,
a human, or other sentient, being] *is*" (198). The Family uses
the human nuclear family as a privileged reference point, a
"power structure" based on the triad of "father-mother-son"
(129). The Sygn (evidently homonymous with "sign") is inspired by semiotics; for it, every structure is characterized
by "the difference between one part of itself and another,"
as well as "the difference between the myriad kinds . . . that
exist on myriad worlds" (135). The Family is always on the
lookout for (mythical) origins, always traced back to the
native soil of Old Earth (117–19), whereas "the Sygn is concerned with preserving the local history of local spaces"
(104). The Family is always "trying to establish the dream
of a classic past . . . in order to achieve cultural stability";
the Sygn is "committed to the living interaction and differ-

ence between each woman and each world from which the right stability and play may flower" (86). The conflict between the Family and the Sygn is therefore one, as Freedman puts it, "between identitarian and nonidentitarian viewpoints" (2000, 159): between an approach that phobically rejects difference and one that actively embraces it.

The Dream of a Common Language. However, this opposition is complicated by the fact that the Family itself, no less than the Sygn, is a loose network, without a center, and not a top-down, hierarchical organization. The "father-mother-son" structure that the Family employs is by no means absolute or rigid. It works as "a model through which to see many different situations. The Family has always been quite loose in applying that system to any given group of humans or nonhumans, breeding or just living together" (129). That is to say, the Family's model of power is more like Lévi-Strauss's "elementary structures of kinship" than it is like *The Donna Reed Show*. Of course, this doesn't make the Family's ideology any the less sexist, racist, homophobic, and authoritarian. But it *does* mean that the Family, just like the Moral Majority or Al-Qaeda, is a thoroughly postmodern phenomenon.[101] As Hardt and Negri put it, speaking of contemporary Islamic radicalism, but in terms that could just as well apply to the Christian Right in America, or to the Family in the novel, "fundamentalism should not be understood as a return to past social forms and values, not even from the perspective of the practitioners...the fundamentalist 'return to tradition' is really a new invention" (148–49). The ostensibly traditional values that such groups invoke "are really directed in reaction to the present social order" (149). It is only *now* that the fundamentalists are able to feel nostalgia for a past that never existed in the first place. Castells similarly argues that Muslim and Christian fundamentalisms, and movements like the right-wing militias in the United States, must be under-

stood as phenomena of globalization and informationaliza-
tion (1997, 12–27) whose entire political practice is organized around decentralized electronic networks and focused
upon such forms of action as interventions in the global
media (68–109). This all suggests that the Family and the
Sygn are not diametrically opposed so much as they are
deeply implicated with one another. They are much like
the "two interpretations of interpretation, of structure, of
sign, of play" discussed by Derrida: "the one seeks to deci-
pher, dreams of deciphering a truth or an origin which escapes play and the order of the sign, and which lives the
necessity of interpretation as an exile. The other, which is
no longer turned toward the origin, affirms play and tries to
pass beyond man and humanism." The first interpretation,
that of the Family, is "saddened, *negative*, nostalgic, guilty,"
as it laments the loss of the center; the second, that of the
Sygn, is an *"affirmation"* that *"determines the noncenter other-*
wise than as loss of the center" (Derrida 292). But even as Derrida highlights the incompatibility between these two positions, he also insists that they must be taken together. They
belong to the same field of possibilities; they are expressions
of the same predicament. Although they are "absolutely ir-
reconcilable," nevertheless "we live them simultaneously
and reconcile them in an obscure economy"; there cannot
be "any question of *choosing*" between them (293). The Family and the Sygn are competing ways of articulating the
local with the universal. But they both take the information
economy for granted. Even the Sygn's admirable politics of
difference presupposes—and remains within the horizon
of—the network society. The Sygn, no less than the Family,
fails to leave space for the singularity of *contact,* which
Delany elsewhere (1999, 123ff) celebrates in contradistinction to the routinized benefits of networking.

Cultural Fugue. Everyone in the universe of *Stars in*
My Pocket Like Grains of Sand is obsessed with an obscure

catastrophe known as Cultural Fugue. A world is said to go into Cultural Fugue when "socioeconomic pressures... reach a point of technological recomplication and perturbation where the population completely destroys all life across the planetary surface" (70). Now, such cataclysmic destruction is relatively rare; over the course of a century, on average only one world out of four hundred will actually fall victim to Cultural Fugue.[102] Nevertheless, whenever a planet is beset with ethnic, political, and economic strife, "if you talk to anyone in the middle of it, among the first things they'll want to know is if their world has gone into Cultural Fugue" (70). Even the rivalry between the Family and the Sygn mostly comes down to "their differing methods of preventing Cultural Fugue" (87). Each side claims that it offers the superior way of stabilizing a troubled world, thereby preserving it from cataclysmic self-destruction. And whenever anyone hears about a world that has been destroyed, or even one that just seems to be having difficulties, the first question that they are likely to ask is whether that world was aligned "with the Family or the Sygn" (71, 91). The irony is that, in actuality, the Family/Sygn conflict occurs "largely on worlds (as more than one commentator has noted) where [Cultural Fugue] wasn't very likely to happen anyway" (87). But this doesn't matter; the threat of self-destruction is palpable to everyone, even if the event never materializes. The danger is part of the atmosphere. The apocalyptic prospect (however improbable) of Cultural Fugue seems to be—as much as the Web, or the information form itself—a defining condition of life in the network society.

Worklessness. It may be rare, but Cultural Fugue *does* happen. *Stars in My Pocket Like Grains of Sand* begins with the catastrophic destruction of a world called Rhyonon; it ends with another world, Velm, barely averting a similar fate. What's especially confounding about Fugue situations is how little they have to do with concrete socioeconomic conditions or with the political conflicts of Family and Sygn. A world can face "a real, desperate, and life-destroying catastrophe," in which "the violence, the death, the anguish" are "immense," and for all that not be anywhere close to Cultural Fugue (70). Conversely, a Fugue situation can materialize suddenly, even on a world or in a culture that is relatively peaceful, prosperous, free, and egalitarian. This is what happens when Rat Korga, the survivor of Cultural Fugue on Rhyonon, visits the city of Morgre on Velm. He becomes an object of fascination for everyone. Crowds spontaneously form and start following him around. In less than a day, all normal routines and economic activities have been abandoned. Instead, everybody mills about aimlessly, hoping just to catch a glimpse of Rat. What is it like to lose an entire world, to lose the only context you ever knew, the one that gave meaning to everything? It is literally impossible to imagine such a thing. And that is why "the whole of Morgre has developed an appetite" for Rat Korga, why everyone is "famished for a taste of [his] survival" (302). The people following Rat are strangely passive, unable to act. They seem to be waiting, as if in suspension, for some incipient event whose nature they cannot picture or imagine. This is a premonitory sign of Cultural Fugue, as the officials of the Web are well aware (367). Indeed, the uncanny *worklessness*[103] of the crowd, its voracious appetite without a real object, is already itself a manifestation of Cultural Fugue: "violence leading to no end . . . not so much destruction ending in death, but rather the perpetual and unremitting destruction of both nature and intelligence run wild and without focus, where anything so trivial and natural as either death or birth is irrelevant" (373). The thing to notice here is the absence of direction, the lack of even a nihilistic or purely negative goal. This is not pure destruction, but destruction all the more disturbing in that it is impure and unfocused. What is most troubling about Cultural Fugue, in other words, is not annihilation per se, so much as the lack of finality, the *meaninglessness* and gratuitousness, of this annihilation. Blanchot calls it the "fear of seeing *here* collapse into the unfathomable nowhere" (259). That is what makes Cultural Fugue so truly unsettling; as Nietzsche puts it, "what really arouses indignation against suffering is not suffering as such but the senselessness of suffering" (1969, 68). Cultural Fugue troubles the imaginations of everyone in the Federation of Habitable Worlds, far beyond its actual probability of occurrence, precisely because it seems so unmotivated, so sudden and senseless. This is what makes it an object of fascination as well as dread.

Reverse Causality. It may be that the arbitrariness of Cultural Fugue, its failure to be explicable and predictable, is itself a consequence of the organization of the network society. When "the Universe is overdetermined," and every person, object, and point in space is a node in the network, then linear causality no longer obtains. Strange saltations and transformations become not merely possible, but inevitable. In the terms of chaos theory, Cultural Fugue is a strange attractor. Its dynamics are abrupt and nonlinear. The process is triggered when a certain threshold or bifurcation point is reached, causing the entire network to flip instantaneously into another state. Cultural Fugue is then a crisis of indetermination, but a massively overdetermined one. Its chaotic interminability is an emergent property of the very fact that everything in the universe has been quantified and digitized and translated and networked, all in the form of information. Deleuze and Guattari write of "reverse causalities that are *without finality* but testify nonetheless to an action of the future on the present, or of the present on the past ... what does not yet exist is already in action, in a different form from that of its existence" (1987, 431). Cultural Fugue, in its shadowy, "irreducible contingency" (Deleuze and Guattari, 1987, 431), is precisely such a future, potential force. And that is why it always has to be warded off, as Deleuze and Guattari say, even before it has manifested itself. Cultural Fugue haunts the network society, precisely because it exists *nowhere* within it: which is to say, it exists— or better, it *insists* or *subsists*—in the in-between, in the interstices of the network, in the nonplaces that its universal equivalences cannot reach.

The Lineaments of Gratified Desire. The threat of Cultural Fugue on Velm seems to be a consequence, not just of Rat Korga's charisma as the survivor of Fugue on another world, but even more of the love[104] between Rat and his host Marq Dyeth (the narrator of most of the novel). Rat and Marq are each other's "perfect erotic object" (179), even though they have never met before. When they are finally brought together on Velm, their consummated and gratified desire is too intense for words. There is nothing transgressive about this desire, nothing in it that suggests prohibition or lack: for in the culture of Velm—unlike those of Rhyonon or the contemporary United States—sex between men is fully recognized and accepted. Nevertheless, Rat's and Marq's erotic encounter exceeds any possibility of recognition, for what they experience over the course of their first day together cannot be measured or compared to anything else. Marq is told by an operative of the Web that his desire is "the most disrupting and random of factors in a very complex equation" (348). But how can such an unquantifiable "factor" be captured in an "equation" at all? The singular intensity of Marq's desire has no counterpart among the universal equivalences of the Web. Such a desire seems capable of overturning a world, of rupturing whatever coherence or equilibrium it once possessed. Which is why, in order to save Velm from Cultural Fugue, the Web forcibly separates Rat from Marq. The people of Morgre stop their aimless milling about; the world returns to its usual, predictable course, and Marq is left behind to mourn. He suffers, even to the point of oblivion, from "the intensity of loss, the absolute vanishing of the possible" (353). And, in the astonishing Epilogue to the novel, he meditates on "the vagaries of translation" (360) and the impossibility of finding a way to translate (or mediate) between his own wounded subjectivity and the lineaments of a universe from which Rat Korga is (for him, at least) absent. "Desire isn't appeased by its object," he says, "only irritated into something more than desire that can join with the stars to inform the chaotic heavens with sense" (370). For Marq's world is not composed of "causes" and "correlations," but of a particular sort of *information:* "information beautiful yet useless to anyone but me, or someone like me, information with an appetite at its base as all information has, yet information to confound the Web and not to be found in any of its informative archives" (369). Such "sense," created by Marq's "more than desire," is not anything that the Web can acknowledge. For it is a Coherence, to paraphrase Deleuze paraphrasing Klossowski, that does not allow the coherence of the world to subsist (Deleuze 1990, 300). And such "information," unlike that of the Web, cannot be recorded or disseminated or exchanged. It cannot be codified in bits or translated according to the criteria of the Web's system of General Information. It cannot be conceptualized or imagined, but only materially *felt*, as rapture or as loss, in the concrete particulars of experience. *Splendor and Misery of Bodies, of Cities.*[105] Marq's and Rat Korga's desire is *extrabeing*, something that has no place in the totality of a world of information. It is as extravagant and beyond representation, as impalpable and yet as overwhelming in its presence, in its sheer materiality, as Cultural Fugue itself. Which may be why the two seem so strangely intertwined and cannot be completely disentangled from one another. Cultural Fugue is the uncanny double, the dark shadow, the destructive inversion, of gratified desire. In excluding the possibility of the one, the network society cannot help but call forth the other.[106]

I Fought the Law, and the Law Won. It's already too late. There is really no way out of all this mess. Or rather, the only way out is the way through. *The street finds its own uses for things,* according to William Gibson's oft-cited maxim. This means that every mechanism of control can be perverted—or *détourned,* as the Situationists say—to new uses. As technologies trickle down to the street, they spread out, multiply their effects, and move in unanticipated directions. Deleuze likes to say, expounding on Spinoza, that *"we do not even know what a body can do..."* (1988b, 17–18). The *potential* of a body, of a technology, or of a machine is always far greater than its official function or its ostensible purpose. The only thing that we can expect, therefore, is the unexpected, as it emerges from repeated processes of dissemination, sampling, and mutation. This is the optimistic, even utopian, view of postmodern politics and aesthetics. But Gibson also hints at a much bleaker vision. In *Neuromancer,* he suggests that "wild," alternative uses of technology are only tolerated because "burgeoning technologies require outlaw zones"; for innovation to take place, there has to be "a deliberately unsupervised playground for technology itself" (11). This means that technologies trickle up, rather than trickle down. The multiple potential uses of a new invention come first; all kinds of crazy possibilities are explored in the outlaw zones. Collective collaboration and cutthroat competition both flourish. But once an invention's possibility space has been thoroughly mapped out, all its dangerous (or insufficiently profitable) potentialities are pruned away so that only a restricted, normalized usage remains. The multinationals grab the technology for themselves, enforcing their patents, copyrights, and trademarks, and closing down unsupervised uses. The "pirate utopia" of freewheeling experimentation gives way to a carefully ordered regime of capital accumulation. Mega-corporations, like K. W. Jeter's DynaZauber, are thereby able to gather all the products of social innovation in order to sell them back to us at a profit. And that is the point at which we are all connected, unable to live without the network. "The only ones who really believed connecting was an unalloyed good thing," Jeter says, "were people who had something to sell and rapists, two categories that weren't that far apart in this world" (175).

What It Means to Live in the Network Society.
So this is what it means to live in the network society. We have moved out of time and into space. Anything you want is yours for the asking. You can get it right here and right now. All you have to do is pay the price. First of all, you must pay the monetary price, since money is the universal equivalent for all commodities. But you also have to pay the *informational price*, since information is also a universal equivalent. Information is the common measure and the medium of exchange for all knowledge, all perception, all passion, and all desire. The universal equivalent for experience, in short. In the network society, experience will be digital or not at all. But this also means that what you get is never quite what you paid for. It's always just a tiny bit less. The mystery of the extraction of surplus value, unveiled by Marx in the context of nineteenth-century capitalism, applies to the information economy as well. The one real innovation of the network society is this: now surplus extraction is at the center of consumption as well as production. When you buy something from Microsoft, or Dyna-Zauber, all the formal conditions of equal exchange are met. And yet there is always something extra, something left over, something that is missing from your side of the equation. A surplus has leaked out of the exchange process. What's missing is what is *more than information:* the qualitative dimension of experience or the continuum of analog space in between all those ones and zeroes. From a certain point of view, of course, this surplus is nothing at all. It is empty and insubstantial, almost by definition. For if it did

exist, it could easily be coded, quantified, and informatized, to any desired degree of accuracy. It is not that there is some hidden essence, basic to human existence, that somehow cannot be rendered by information machines. It is rather that information can all too well account for everything; there is literally nothing that it cannot capture and code. But this *nothing* is precisely the point. Because of this nothing, too much is never enough, and our desires are never satisfied. This *nothing* insinuates itself into our dreams. It is what always keeps us coming back for more. And *that* is "the dirty little secret that corporations know."

The Theme Park of Desires. K. W. Jeter puts it best: "You're looking at the future here, pal; the future and the present and the past, all rolled up in one. The goal of commerce is to destroy history, to put its customers in the eternal Now, the big happy theme park of desires that are always at the brink of satisfaction but somehow never get there. Because if they did, the game would be over and everybody would go home" (380).

Connected. You may say that all this is merely science fiction. None of it is happening: not now, not here, not yet. But science fiction does not claim to be reportage, just as it does not claim to be prophecy. It does not actually represent the present, just as it does not really predict the future. Rather, it involves both the present and the future, while being reducible to neither. For science fiction is about the shadow that the future casts upon the present. It shows us how profoundly we are *haunted* by the ghosts of what has not yet happened. This is the condition that K. W. Jeter describes for us, in his account of the network society: "The little machines continued their work, visibly, like some nightmare of a future that had already arrived" (17).

Notes

1. For a celebratory history of these developments, from cybernetics to complexity theory and beyond, see Kelly. For a more nuanced and critical account, see Hayles.

2. To use Schumpeter's term for capitalism's reign of continual, ruthless innovation.

3. This is the overall argument of Michel Foucault's *Archaeology of Knowledge*. See also Foucault's essay on schizophrenic linguistics, *Sept propos sur les septième ange*.

4. To borrow a metaphor from Rudolph Gasché.

5. I thank Friedrich Kittler for calling this to my attention.

6. Deleuze often compares the categorical imperative of Kant's Second Critique with the execution machine that inscribes the text of the Law into the very flesh of the transgressor in Kafka's "In the Penal Colony." See Deleuze (1984, x–xi) and Deleuze and Sacher-Masoch (83–84).

7. As David Bowie's alien billionaire does in Nicholas Roeg's prescient film *The Man Who Fell to Earth*.

8. "Tout, au monde, existe pour aboutir à un livre" (378).

9. As rendered in Jorge Luis Borges's short essay "Paschal's Sphere" (351–53).

10. See the SCP Web site, http://www.notbored.org/scp-performances.html. Accessed April 2001.

11. Steven Levy gives a laudatory account of the Cypherpunk movement and its battles with the U.S. government over encryption policy. For a critique of Cyberpunk culture and ideology, see Borsook (73–114).

12. Thus the investor is to the market as Proust's narrator is to Albertine.

13. I borrow this term from Manovich (45ff).

14. As noted by Nicole Maersch on the CDNow Web site, http://www.cdnow.com. Accessed April 2001.

15. See Sun Ra's film *Space is the Place* for an explanation of this process.

16. For a summary of current "fair use" guidelines, see CETUS (Consortium for Educational Technology for University Systems),

"Fair Use of Copyrighted Works," available online, together with a list of useful links, at http://www.cetus.org/fairindex.html. Accessed April 2001.

17. I lack the space here to go into an extended discussion of how the DMCA's anticircumvention provisions can limit access so severely that questions of fair use effectively become moot. But see, for instance, the statement of major library associations, available online at http://www.arl.org/info/letters/dmca_80400.html. Accessed April 2001.

18. The science fiction writer Harlan Ellison filed a lawsuit in 2001 to suppress the piracy of some of his short stories on Usenet. See "Harlan Ellison Fights for Creators' Rights," http://harlanellison.com/KICK/kick_rls.htm.

19. http://detritus.net.illegalart/beck/. Accessed April 2001.

20. No lawsuit was ever actually filed, due in part to the difficulty of tracking down Illegal Art or even finding out who stood behind it. Correspondence between Illegal Art and various copyright lawyers is available at http://www.rtmark.com/lawletters.html. Accessed April 2001.

21. My discussion of *Deconstructing Beck* is adapted from my article about the album.

22. I discuss this process in greater detail in my essay "Supa Dupa Fly: Black Women as Cyborgs in Recent Hip-hop Videos" (to appear as a chapter of my forthcoming book, *Beauty Lies in the Eye*).

23. Len Bracken, who made an earlier translation of *The Society of the Spectacle*, argues with justice that Nicholson-Smith's translation is inaccurate and misleading. In Bracken's translation, the lines I am quoting read: "All that was directly lived has distanced itself in representation... The images that are detached from every aspect of life flow into a common stream where the unity of life can never be reestablished." See http://www.lenbracken.com/. Accessed April 2001.

24. I develop the argument of this paragraph at much greater length in my essay "The Life (after Death) of Postmodern Emotions" (to appear as a chapter in my forthcoming book, *Beauty Lies in the Eye*).

25. This paragraph is adapted from discussion of *Glamorama* in my online book *Stranded in the Jungle*. See http://www.shaviro.com/Stranded/34.html. Accessed April 2001.

26. I selected this article almost at random, as an expression of common ideas about celebrity culture.

27. I discuss Warhol's Marilyn portraits in "The Life (after Death) of Postmodern Emotions."

28. As I write these lines (2001), the Trojan Room Coffee Machine is still running at http://www.cl.cam.ac.uk/coffee/coffee.html. Accessed April 2001.

29. http://www.jennicam.org. Accessed April 2001.

30. See "Jennifer Ringley's Diary" for 12 August 2000, in the "Journal: Archives" section of the JenniCam Web site.

31. http://www.crime.com/info/jailcam_redirect.html. Accessed April 2001.

32. See Platt, and Corrin. The Web site itself is at http://weliveinpublic.com. Accessed April 2001.

33. http://www.kathicam.com. Accessed April 2001.

34. For example, in April 2001 one could find sites like http://mija.nu/.

35. For example, in April 2001 one could find sites like http://www.hush-hush.com/1stflirt/index01.htm.

36. For whatever reason, a Google search on the terms "porno webcam" turned up mostly German- and French-language sites. (Search conducted on 5 April 2001.)

37. http://www.stuart-tiros.com. Accessed April 2001.

38. William Scarbrough, *Prosthetic* (Michael Gold Gallery, New York, 22 April–28 May 1999). Now available on CD-ROM.

39. My discussion of Stuart Tiros and *Prosthetic* is adapted from my article about the installation (1999a).

40. Rorty's *Philosophy and the Mirror of Nature* is a popular account of how thinkers as diverse as Wittgenstein, Heidegger, and Dewey, in the first half of the twentieth century, and Quine, Sellars, Davidson, Derrida, Foucault, and many others, in the second half, all converge in their rejection of representationalist assumptions.

41. More precisely: cognitive scientists start from the assumption that minds, like computers, are "information processors" that work by "applying certain operations to mental content. This mental content has to be about something, and this means that we have to represent in our minds (and ultimately, in our brains) the things that mental content are about." Thus the "constitutive problem of Cognitive Science" is "the question of how things (and processes) outside minds can be represented inside minds (or

inside machines, for that matter) so that minds (or machines) can 'do' things with them" (Ortony). Though recent AI research has put more emphasis on distributed processes and embodied inter- actions (Brooks), representationalist assumptions remain basic to the field.

42. The screenplay for *The Matrix* is available online at http:// www.geocities.com/Area51/Capsule/8448/Matrix.txt. Accessed April 2001.

43. My comments here on *The Matrix* may be taken as the rudiments of a "Kantian" reading of the film, dealing with the unavoidable *paralogisms* and *antimonies* into which it falls—this in deliberate opposition to the "Hegelian" reading offered by Žižek, which turns explicitly on Hegel's critique of Kant.

44. Nietzsche (1968a) actually writes: "I fear we are not getting rid of God because we still believe in grammar" (38).

45. The figure of Palmer Eldritch (Dick 1991a) seems especially to take on this role, among other characters in Dick's novels.

46. Most notably, *Valis*.

47. This should not be taken to imply that I am endorsing a "metaphysical" reading of Dick, as opposed to a materialist or Marxist one that would emphasize his concerns with commodity culture. My foregrounding of Dick's ontological concerns over epistemological and hermeneutical ones is entirely separate from the question as to how much Dick's vision is social, political, and material, and how much is metaphysical and religious. If any- thing, I would want to say that Dick's writing expresses what Foucault calls "an incorporeal materialism" (1972, 231).

48. This paragraph is adapted from my discussion of "All Is Full of Love" in my online book *Stranded in the Jungle*. I discuss the video at much greater length in my essay "The Erotic Life of Machines."

49. *Dancing with the Virtual Dervish* was originally presented at the Banff Centre for the Arts in 1993 by Diane Gromala in collab- oration with Yakov Sharir.

50. My discussion of *Dancing with the Virtual Dervish* is adapted from my article about the piece.

51. See especially Ballard's *Crash* and *The Atrocity Exhibition*.

52. Cf. Kant: "Thoughts without content are empty; intuitions without concepts are blind" (1996, 107).

53. See especially the Web site "Astraea's Multiplicity Resources

and Controversies" (http://astraeasweb.net/plural/. Accessed April 2001).

54. Jerry Fodor distinguishes between the question as to "whether minds are 'Turing equivalent' (that is: whether there is anything that minds can do that Turing machines can't," and the question as to "whether the architecture of (human) cognition is interestingly like the architecture of Turing's kind of computer." He points out that "the answer to the second could be 'no' or 'only in part' whatever the answer to the first turns out to be" (105).

55. http://tuxedo.org/~esr/jargon/html/. Accessed April 2001.

56. McKenna, just like Vinge and Kurzweil, makes his prediction by observing that the rate of technological change is continually accelerating, and therefore calculating that, at a certain point, the tangent of the curve of change effectively becomes infinite.

57. For a discussion of the strange, delirious logic of derivatives in the global financial marketplace, see Doyle (2003).

58. For example: "We are like water creatures looking up at the land and air and wondering how we can survive in that alien medium. The water we live in is Time. That alien medium we glimpse beyond is Space. And that is where we are going" (Burroughs 1983, 41).

59. As in Abel Ferrara's film *New Rose Hotel*, discussed later, which takes place almost entirely in the third layer of the space of flows, and mostly in various hotel rooms. This allows Ferrara to use New York as a stand-in for Berlin and Tokyo.

60. Or, in the words of Castells, "pale chamois" (2000b, 447).

61. "Ride the light" is the official corporate slogan of the telecommunications giant Qwest Communications.

62. The allusion is to *The Organization Man* by William H. Whyte, a celebrated analysis of the conformism of American corporate culture in the 1950s.

63. *Rattisage* is police slang, translated as "dragnet operation" (Hérail and Lovatt).

64. "Thus *death* reveals itself as the *ownmost nonrelational possibility not to be bypassed*" (Heidegger 232).

65. Quoted in James (115–16).

66. Jeter has written several novelistic sequels to the *Blade Runner* movie.

67. William Beard, personal communication.

68. This parallel was suggested to me by Bruce Reid (personal communication).

69. My discussion of "tendential fall" draws upon Michael Hardt's reading notes for *Capital*, available online at http://www.duke.edu/~hardt/THREE-3.htm. Accessed April 2001.

70. See http://detroityes.com/index.html. Accessed April 2001.

71. Walt Disney's vision of an "experimental prototype community of the future," originally modeled in the EPCOT Center portion of Disney World, took shape in the Disney-owned planned city of Celebration, Florida. See the official Celebration Web site, http://www.celebrationfl.com/ (accessed April 2001), and Ross.

72. These changes date back to the early nineteenth century. Before the automobile and the telephone, railroads produced new sensations of speed and the telegraph annihilated distance. See Standage, for a discussion of the telegraph, and Crary, for a more general discussion of the mutations of the sensorium under the impact of modernity.

73. For information about the biochemical effects of drugs, see Regan, Pellerin, and Kuhn et al.

74. Jeff Noon makes an explicit parallel between drugs and music, and proposes both as technologies of remixing subjectivity, in his novel *Needle in the Groove*.

75. "Such a fallacious inference will thus have its basis in the nature of human reason, and will carry with it an illusion that is unavoidable although not unresolvable" (Kant 1987, 382).

76. Here, as elsewhere in this book, my approach to Kant is inspired by Deleuze's reading of Nietzsche as a radical neo-Kantian in his books on both thinkers and by Foucault's remarks about "that opening made by Kant in Western philosophy when he articulated, in a manner that is still enigmatic, metaphysical discourse and reflection on the limits of our reason" (1998, 76). If the bulk of *The Critique of Pure Reason* constructs the figure of the "empirico-transcendental doublet," then the "Transcendental Dialectic" section of the book returns to that figure's genesis, in order to articulate the possibilities of its undoing.

77. The psychedelic account of nonrepresentational "knowledge" that I am offering here is of course not precisely what Jameson has in mind. Jameson refers to Althusser's ultimately Spinozan distinction between ideology and science. Subjectivity is unavoidably representational and ideological, but Spinoza and

Althusser claim that an asubjective, "scientific" discourse is possible *sub specie aeternatis*. Psychedelia, like Spinozistic/Althusserian "science," is radically asubjective, nonrepresentational, and ontological, but I certainly do not concur with those who would make extravagant epistemological claims for it.

78. Though this inevitable inadequacy has not stopped me from making my own attempt to give a verbal account of the LSD experience. See the chapter "Acid" in my online book *Stranded in the Jungle*, at http://www.shaviro.com/Stranded/16.html. Accessed April 2001.

79. McKenna characterizes psychedelic experience in terms of "information" over and over again. For instance, earlier in the same essay: "it is as if information were being presented three-dimensionally and deployed fourth-dimensionally, coded as light and as evolving surfaces."

80. For contemporary Marxist theory, the old duality of base and superstructure no longer functions in late capitalism. As Hardt and Negri put it: "Postmodernization and the passage to Empire involve a real convergence of the realms that used to be designated as base and superstructure. Empire takes form when language and communication, or really when immaterial labor and cooperation, become the dominant productive force" (385–86). All this is a consequence of "the real subsumption of society under capital," in contrast to its merely formal subsumption in earlier stages of capitalism.

81. See Deleuze (1988b, 48–51) for a discussion of affect. Strictly speaking, affects are not any more mental than they are physical, but they are related to thought in that they are reflexive, involving the experience of a "passage from one state to another."

82. I am arguing here, not that affect is necessarily conscious and subjective, but that subjectivity and consciousness are necessarily affective, in the sense defined by Spinoza and Deleuze. For the idea of impersonal, asubjective affect, and the role it plays in the production of subjectivity, see Shaviro (2001).

83. See Fausto-Sterling for a nuanced account of the effects of testosterone and other so-called sex hormones, and for a critique of the biological essentialism espoused by Sullivan.

84. According to the researchers, "plants may be actively 'choosing' the species of fungus that supports the highest growth for the plant ... the consequences are the same as if it were a cognitive choice."

85. See Dorion Sagan's Foreword to McMenamin, as well as Margulis and Sagan (253–55). I have not been able to track down Drum's original proposal; I know of it only from these two citations from Sagan.

86. See, for instance, Margulis and Schwartz.

87. I would like to thank Richard Doyle for calling this reference to my attention.

88. Strictly speaking, "sociobiology" has been replaced by "evolutionary psychology" as the discipline that proposes ostensibly Darwinian explanations of human culture and behavior. But the polemics, for and against, are pretty much the same in both instances.

89. For the rest of this paragraph, I only look at Pinker's method of explanation. Dennett's arguments against Gould, and in favor of a self-consciously reductive adaptationism, are far stronger than Pinker's, and would require much more space to discuss adequately. I focus here on Pinker's rhetoric, not in order to get off a few cheap shots, but because I find this rhetoric symptomatic of how problems are posed in the context of network culture.

90. Of course, not much insight on these matters can be expected from someone who attributes the concept of "false consciousness" to Michel Foucault.

91. It is precisely in order to get around this difficulty that Pinker argues that the human mind is massively modular (27–31). There is no such thing as a general mind, but only multiple "mental modules," each of which has a particular function that can be separately understood by "reverse engineering," just as the function of the eye can be. Jerry Fodor cogently argues that this radical atomization of mental processes cannot be right, for it cannot possibly explain how the various mental modules are coordinated and why one of them gets called upon in any particular instance as opposed to another.

92. Discussing Nonzero, Pinker takes exception to Wright's claim that globalization is the ultimate goal of all life and of natural selection, but he agrees with Wright "that human nature . . . put our species on an escalator of cultural and moral progress, culminating in today's globalization" (Pinker and Wright).

93. According to N. Katherine Hayles, homeostasis was important to the "second wave" of cybernetic theory in the 1960s and 1970s (131–59), while the more recent "third wave" of cyber-

netic theory is more interested in dynamic, self-organizing processes (222–46).

94. Strictly speaking, what is evolutionarily stable is a particular mix of strategies, or behavioral patterns, in a given population as a whole.

95. I am perfectly aware that Thornhill and Palmer are not arguing that rape is necessarily an efficient means for a male to propagate his genes today, but only that it was so in the distant past, before the invention of condoms. The rhetorical point of my example is that, even in this form, Thornhill and Palmer's theory is an unintentional reductio ad absurdum of narrowly functionalist explanations.

96. Actually, genes can spread laterally as well. That is what viruses do when they invade cells. It is also how *plasmids* operate: those bits of DNA that are exchanged during "bacterial sex," often conferring traits like immunity to antibiotic drugs (see Margulis and Sagan 85–98, and Brown 251–58). Today, plasmids are increasingly used as "cloning vectors" for gene therapy and genetic engineering (Brown 251–58). Still, inheritance by descent evidently remains the chief mechanism for the transmission of genetic material among eukaryotic organisms.

97. It is important to note that this is not the same thing as the question about *units of selection*—genes or memes—that I discussed earlier.

98. Actually, the narrator of the greater part of the novel, Marq Dyeth, is male. But in Marq's world, and consequently in all sections Marq narrates, the pronoun "she" is used to refer to sentient beings of any gender; "he" is reserved for the object of the speaker's sexual desire. This is one way in which the English-language reader is forced to become aware of the arbitrariness, and limited applicability, of his/her own cultural assumptions, and linguistic and social conventions.

99. It is worth noting that *Stars in My Pocket* was first published in 1984, some six years before the invention of the World Wide Web.

100. For Freedman, the Web exemplifies "positivistic, contemplative, ruling-class knowledge," as opposed to a more dialectical and historically grounded sort of comprehension (2000, 151). While I don't really disagree with this, I find it more useful to think of the Web in structural terms, rather than ideological ones: infor-

mation as universal equivalent, a flow linking otherwise incommensurable cultures and worlds.

101. Thus I disagree with Freedman's claim that the Sygn "is certainly more coherent with the 'modernity' of the epoch" depicted in the novel than is the Family (159).

102. We are told that Cultural Fugue has happened "forty-nine times in the last two hundred eighty years," measuring time on a planet whose "years are a bit longer than Old Earth Standard" (71). If we assume this equals a time span of 300 Earth years, then Cultural Fugue happens about 16.33 times per century. In a galaxy of 6,200 inhabited worlds, the probability of any one planet suffering Cultural Fugue in a century is therefore roughly 0.0026, or about one quarter of one percent. It is important to note that, while the novel allows us to make such calculations, it also warns us that they are not to be trusted, for such calculations give a sense of false precision to concepts whose boundaries are inherently fuzzy. Questions like "what is the exact human population of the universe" (73) and "exactly *how* many survivors were there" from a world that went into Cultural Fugue (155) simply do not have precise and meaningful answers.

103. I am using *worklessness* to translate Blanchot's *désoeuvrement*. Blanchot's translator, Ann Smock, prefers to render this word as either "inertia" or "lack of work" (13), but I stick with the less elegant *worklessness,* taken from earlier translations of Blanchot, because it better suggests a state of aimlessness, or being out of sorts. Note, too, that Blanchot links *worklessness* closely to *fascination.*

104. Or at least potential for love, between two men who have not been granted "time enough to find out" if love might grow (372).

105. As Delany calls his (thus far) uncompleted sequel to *Stars in My Pocket Like Grains of Sand.*

106. This is where my reading of the novel diverges most sharply from Freedman's (2000). For I see Marq's desire as a process of selection, and an affirmation of singularity—and not, as Freedman argues, as an Adornian or Sartrean project of totalization (163–64). Again, Freedman suggests that Marq's very personal invocation of "information" as desire "may, in fact, ultimately be all that stands between survival and Cultural Fugue" (163). But I argue to the contrary that this desire is in fact intimately related to the danger of Cultural Fugue. Both singular desire and Cul-

tural Fugue resist the Web's regime of information as a universal equivalent. The crucial distinction between singular desire and Cultural Fugue is precisely what cannot be articulated in the context of the network society, but requires some opening beyond it.

Bibliography

Adorno, Theodor W. 1991. *The Culture Industry: Selected Essays on Mass Culture*. Ed. J. M. Bernstein. New York: Routledge.

Agamben, Giorgio. 2000. *Means without End: Notes on Politics*. Trans. Vincenzo Binetti and Cesare Casarino. Minneapolis: University of Minnesota Press.

Ashbery, John. 1970. *Some Trees*. New York: Corinth Books.

Attali, Jacques. 1985. *Noise: The Political Economy of Music*. Trans. Brian Massumi. Minneapolis: University of Minnesota Press.

Bagemihl, Bruce. 1999. *Biological Exuberance: Animal Homosexuality and Natural Diversity*. New York: St. Martin's Press.

Ballard, J. G. 1985. *Crash*. New York: Vintage.

———. 1990. *The Atrocity Exhibition*, Rev. ed. San Francisco: ReSearch.

———. 2001. *Super-Cannes*. London: Picador.

Barlow, John Perry. 1994. "The Economy of Ideas." *Wired*, March. www.siteofthesentient.com/barlow.html. Accessed April 2001.

———. 1996. "A Declaration of the Independence of Cyberspace." www.eff.org/pub/Misc/Publications/John_Perry_Barlow/barlow_0296.declaration. Accessed April 2001.

Barthes, Roland. 1989. *The Rustle of Language*. Trans. Richard Howard. Berkeley: University of California Press.

Bataille, Georges. 1985. *Visions of Excess*. Trans. Allan Stoekel, with Carl R. Lovitt and Donald M. Leslie Jr. Minneapolis: University of Minnesota Press.

———. 1988. *The Accursed Share*, Vol. 1. Trans. Robert Hurley. New York: Zone Books.

Bateson, Gregory. 2000. *Steps to an Ecology of Mind: Collected Essays in Anthropology, Psychiatry, Evolution, and Epistemology*. Chicago: University of Chicago Press.

Baudrillard, Jean. 1983. *In the Shadow of the Silent Majorities*. Trans. Paul Foss, John Johnston, and Paul Patton. New York: Semiotext(e).

———. 1988. *The Ecstasy of Communication*. Trans. Bernard Schutze and Caroline Schutze. New York: Semiotext(e).

———. 2001. *Selected Writings,* 2d ed. Ed. Mark Poster. Trans. Paul Foss, Paul Patton, and Philip Beitchman. Stanford, Calif.: Stanford University Press.

Beck. 1996. *Odelay.* Geffen Records.

———. 1999. *Midnite Vultures.* Interscope Records.

Benjamin, Walter. 1969. *Illuminations.* Ed. Hannah Arendt. Trans. Harry Zohn. New York: Schocken Books.

———. 1999. *The Arcades Project.* Trans. Howard Eiland and Kevin McLaughlin. Cambridge, Mass.: Belknap Press, Harvard University Press.

Berkeley, George. 1996. *Philosophical Works.* New York: Everyman.

Bey, Hakim. 1991. *T. A. Z.: The Temporary Autonomous Zone, Ontological Anarchy, Poetic Terrorism.* New York: Autonomedia. www.hermetic.com/bey/taz_cont.html. Accessed April 2001.

Blackmore, Susan. 1999. *The Meme Machine.* New York: Oxford University Press.

Blanchot, Maurice. 1982. *The Space of Literature.* Trans. Ann Smock. Lincoln: University of Nebraska Press.

Bloom, Harold. 1992. *The American Religion.* New York: Simon and Schuster.

———. 1997. *The Anxiety of Influence,* 2d ed. New York: Oxford University Press.

Borges, Jorge Luis. 1999. *Selected Non-Fictions.* Ed. Eliot Weinberger. Trans. Esther Allen, Suzanne Jill Levine, and Eliot Weinberger. New York: Viking.

Borsook, Paulina. 2000. *Cyberselfish.* New York: Public Affairs.

Bowles, Jane. 1978. *My Sister's Hand in Mine: The Collected Works of Jane Bowles.* New York: The Ecco Press.

Brin, David. 1998. *The Transparent Society.* New York: Perseus.

Brooks, Rodney. 2002. *Flesh and Machines: How Robots Will Change Us.* New York: Pantheon Books.

Brown, T. A. 1992. *Genetics: A Molecular Approach,* 2d ed. New York: Chapman and Hall.

Burroughs, William S. 1979. *Ah Pook Is Here.* London: John Calder.

———. 1981. *Cities of the Red Night.* New York: Henry Holt.

———. 1983. *The Place of Dead Roads.* New York: Holt, Reinhart, and Winston.

———. 1987. *The Western Lands.* New York: Penguin.

———. 1992a. *Naked Lunch.* New York: Grove.

———. 1992b. *Nova Express.* New York: Grove.

Burroughs, William S., and Brion Gysin. 1978. *The Third Mind.* New York: Viking.

Bush, George W. 2000. "Gore-Bush Second Presidential Debate, October 11, 2000, Winston-Salem, North Carolina." www.cbsnews.com/now/story/0,1597,240443-412,00.shtml. Accessed April 2001.

Butler, Octavia. 1995. *Parable of the Sower.* New York: Warner Books.

———. 1996. *Bloodchild and Other Stories.* New York: Seven Stories Press.

Cadigan, Pat. 2001. *Synners.* New York: Four Walls Eight Windows.

Castells, Manuel. 1997. *The Power of Identity.* Vol. 2 of *The Information Age: Economy, Society, and Culture.* Malden, Mass.: Blackwell.

———. 2000a. *End of Millennium.* Vol. 3 of *The Information Age: Economy, Society, and Culture,* 2d ed. Malden, Mass.: Blackwell.

———. 2000b. *The Rise of the Network Society.* Vol. 1 of *The Information Age: Economy, Society, and Culture,* 2d ed. Malden, Mass.: Blackwell.

———. 2001. *The Internet Galaxy: Reflections on the Internet, Business, and Society.* New York: Oxford University Press.

Chion, Michel. 1994. *Audio-Vision.* Trans. Claudia Gorbman. New York: Columbia University Press.

Clark, Andy. 2001. *Mindware: An Introduction to the Philosophy of Cognitive Science.* New York: Oxford University Press.

Comaroff, Jean, and John Comaroff. 1999. "Alien Nation: Zombies, Immigrants, and Millenial Capitalism." *CODFESRIA Bulletin* 3–4: 17–26. history.wisc.edu/bernault/magical/comaroff%20text.pdf. Accessed April 2001.

Conlin, Michelle. 2000. "And Now, the Just-in-Time Employee." *Business Week Online,* 28 August. www.businessweek.com/2000/00_35/b3696044.htm. Accessed April 2001.

Corrin, Tanya. 2001. "The Harris Experiment." *Fucked Company Forum.* forum.fuckedcompany.com/ubb/Forum5/HTML/003321.html. Accessed April 2001.

Crary, Jonathan. 1992. *Techniques of the Observer.* Cambridge: MIT Press.

Dantec, Maurice G. 1999. *Babylon Babies.* Paris: Gallimard Folio.

Darwin, Charles. 1998. *The Origin of Species.* New York: Modern Library.

Davis, Allison. 1998. "Plants Get the Message, Too." *National Institute of Health News Advisory.* 11 November. www.nih.gov/news/pr/nov98/nigms-12.htm. Accessed April 2001.

Davis, Erik. 2000. Take the Red Pill. *Feed,* 14 August. www.techgnosis.com/redpill1.html. Accessed April 2001.

Dawkins, Richard. 1982. *The Extended Phenotype.* New York: Oxford University Press.

———. 1987. *The Blind Watchmaker.* New York: W. W. Norton and Company.

———. 1989. *The Selfish Gene.* 2d ed. New York: Oxford University Press.

———. 1991. "Viruses of the Mind." www.santafe.edu/~shalizi/Dawkins/viruses-of-the-mind.html. Accessed April 2001.

———. 1998. *Unweaving the Rainbow.* New York: Houghton Mifflin.

De La Soul. 1989. *3 Feet High and Rising.* Tommy Boy Recordings.

Debord, Guy. 1994. *The Society of the Spectacle.* Trans. Donald Nicholson-Smith. New York: Zone Books.

Delany, Samuel R. 1985. *Stars in My Pocket Like Grains of Sand.* New York: Bantam Spectra.

———. 1999. *Times Square Red, Times Square Blue.* New York: New York University Press.

Deleuze, Gilles. 1983. *Nietzsche and Philosophy.* Trans. Hugh Tomlinson. New York: Columbia University Press.

———. 1984. *Kant's Critical Philosophy.* Trans. Hugh Tomlinson and Barbara Habberjam. Minneapolis: University of Minnesota Press.

———. 1988a. *Foucault.* Trans. Seán Hand. Minneapolis: University of Minnesota Press.

———. 1988b. *Spinoza: Practical Philosophy.* Trans. Robert Hurley. San Francisco: City Lights Books.

———. 1990. *The Logic of Sense.* Trans. Mark Lester. New York: Columbia University Press.

———. 1993. *The Fold: Leibniz and the Baroque.* Trans. Tom Conley. Minneapolis: University of Minnesota Press.

———. 1994. *Difference and Repetition.* Trans. Paul Patton. New York: Columbia University Press.

———. 1995. *Negotiations.* Trans. Martin Joughin. New York: Columbia University Press.

Deleuze, Gilles, and Félix Guattari. 1983. *Anti-Oedipus.* Trans. Robert Hurley, Mark Seem, and Helen R. Lane. Minneapolis: University of Minnesota Press.

————. 1987. *A Thousand Plateaus.* Trans. Brian Massumi. Minneapolis: University of Minnesota Press.

Deleuze, Gilles, and Leopold von Sacher-Masoch. 1989. *Masochism: Coldness and Cruelty and Venus in Furs.* Trans. Jean McNeill. New York: Zone Books.

Deltron 3030. 2000. *Deltron 3030.* 75Ark.

Dennett, Daniel. 1995. *Darwin's Dangerous Idea.* New York: Simon and Schuster.

————. 1998. *Brainchildren (Representation and Mind).* Cambridge: MIT Press.

Derrida, Jacques. 1980. *Writing and Difference.* Trans. Alan Bass. Chicago: University of Chicago Press.

Descartes, René. 1984. *The Philosophical Writings of Descartes.* Vol. 2. Trans. John Cottingham, Robert Stoothoff, and Dugald Murdoch. New York: Cambridge University Press.

Di Filippo, Paul. 1997. *Ciphers.* Campbell, Calif.: Cambrian Publications, and San Francisco: Permeable Press.

————. 1998. *Ribofunk.* New York: Avon Press.

Dibbell, Julian. 2001. "Pirate Utopias." *Feed,* 20 February. www.feedmag.com/templates/default.php3?a_id=1624. Accessed April 2001.

Dick, Philip K. 1981. *Valis.* New York: Bantam.

————. 1991a. *The Three Stigmata of Palmer Eldritch.* New York: Vintage.

————. 1991b. *Ubik.* New York: Vintage.

Divjak, Carol. 2000. "Free Speech—Singapore Style." *World Socialist Web Site,* 20 June. www.wsws.org/articles/2000/jun2000/sing-j20.shtml. Accessed April 2001.

Doyle, Richard. 2002. "LSDNA: Consciousness Expansion and the Emergence of Biotechnology." In *Semiotic Flesh: Information and the Human Body,* ed. P. Thurtle and R. Mitchell. Seattle: University of Washington Press.

————. 2003. *Wetwares: Experiments in Postvital Living.* Minneapolis: University of Minnesota Press.

Dr. Dre. 1999. "Still D.R.E." On *Dr. Dre 2001.* Interscope Records.

Drudge, Matt. 1997. "Gore to Dine at Gates' Techno Mansion." *Wired News,* 8 May. www.wired.com/news/print/0,1294,3719,00.html. Accessed April 2001.

Ehrenberg, Rachel, and Nancy Ross-Flanigan. 2001. "Picky Plants: Do They 'Choose' the Best Fungal Partner?" *University of Mich-*

igan News and Information Services News Release, 30 August. www.umich.edu/~newsinfo/Releases/2001/Aug01/r080301d. html. Accessed August 2001.

Elliott, Missy. 1997. "The Rain (Supa Dupa Fly)." On *Supa Dupa Fly*. Elektra/Asylum.

Elliott, Stuart. 2001. "Advertising: ESPN Uses a 'Hostile' Web Site." *The New York Times*, 27 February. www.nytimes.com/2001/02/27/technology/27ADCO.html. Accessed April 2001.

Ellis, Brett Easton. 2000. *Glamorama*. New York: Vintage.

Ellis, Warren, and Darick Robertson. 1998a. *Back on the Street*. Vol. 1 of *Transmetropolitan*. New York: DC.

———. 1998b. *Lust for Life*, Vol. 2 of *Transmetropolitan*. New York: DC.

———. 2000. *The New Scum*. Vol. 4 of *Transmetropolitan*. New York: DC.

———. 2002. *Gouge Away*. Vol. 6 of *Transmetropolitan*. New York: DC.

———. 2003. *Dirge*. Vol. 8 of *Transmetropolitan*. New York: DC.

Eshun, Kodwo. 1998. *More Brilliant Than the Sun: Adventures in Sonic Fiction*. London: Quartet Books.

Experience Music Project. 2000. *Fact Sheet*. www.emplive.com/visit\press_room/fact_sheet.pdf. Accessed April 2001.

Experimental Interaction Unit. 1998. *Inevitable Dilemmas of the Human Condition*. www.eiu.org/experiments/i-bomb/.

Farber, Phillip H. 1993. "Mushrooms, Sex and Society: An Interview with Terence McKenna." *New History*, June. members.aol.com/discord23/mckenna.htm. Accessed April 2001.

Fausto-Sterling, Anne. 2000. *Sexing the Body: Gender Politics and the Construction of Sexuality*. New York: Basic Books.

Fodor, Jerry. 2000. *The Mind Doesn't Work That Way*. Cambridge: MIT Press.

Folkers, Richard. 1997. "Xanadu 2.0: Bill Gates's Stately Pleasure Dome and Futuristic Home." *US News and World Report*, 1 December. www.usnews.com/usnews/issue/971201/1gate.htm. Accessed April 2001.

Foucault, Michel. 1970. *The Order of Things: An Archaeology of the Human Sciences*. New York: Vintage.

———. 1972. *The Archaeology of Knowledge*. Trans. A. M. Sheridan Smith. New York: Pantheon.

———. 1978. *The History of Sexuality: An Introduction*. Vol. 1 of *The History of Sexuality*. Trans. Robert Hurley. New York: Pantheon.

———. 1979. *Discipline and Punish.* Trans. Alan Sheridan. New York: Vintage.

———. 1983. "The Subject and Power." In *Michel Foucault: Beyond Structuralism and Hermeneutics,* 2d ed. Ed. Hubert L. Dreyfus and Paul Rabinow. Chicago: University of Chicago Press.

———. 1986a. *Sept propos sur le septième ange.* Paris: Fata Morgana.

———. 1986b. *The Use of Pleasure.* Vol. 2 of *The History of Sexuality.* Trans. Robert Hurley. New York: Vintage.

———. 1998. *Aesthetics, Method, and Epistemology.* Vol. 2 of *Essential Works of Foucault.* Trans. Robert Hurley et al. New York: The New Press.

Fox Market Wire. 2001. "Nortel Introduces Network Technology That Can Track Web Use by Individuals," 31 January.

Freedman, Carl. 1984. "Towards a Theory of Paranoia: The Science Fiction of Philip K. Dick." *Science Fiction Studies* 11 (March): 15–24.

———. 2000. *Critical Theory and Science Fiction.* Hanover: Wesleyan University Press.

Friedman, David. 1996. *Hidden Order: The Economics of Everyday Life.* New York: HarperBusiness.

Funke, Rainier. 1995. "Cyberspace versus homunkulus? Beginnt das problem im virtuellen?" www.design.fh-potsdam.de/FB4/Publikation/1995/Funke_Cyberspace/F%unke_Cyberspace.html. Accessed April 2001.

García Márquez, Gabriel. 1998. *One Hundred Years of Solitude.* Trans. Gregory Rabassa. New York: Harper Perennial Library.

Gasché, Rudolph. 1988. *The Tain of the Mirror.* Cambridge: Harvard University Press.

Gates, Bill. 2000. *Business @ the Speed of Thought: Succeeding in the Digital Economy.* New York: Warner Books.

Gehr, Richard. 1992. "Omega Man: A Portrait of Terence McKenna." *Richard Gehr's Rubrics and Tendrils.* 5 April. www.levity.com/rubric/mckenna.html. Accessed April 2001.

Gibson, William. 1984. *Neuromancer.* New York: Ace Books.

———. 1986. *Count Zero.* New York: Ace Books.

———. 1987. *Burning Chrome.* New York: Ace Books.

———. 2000. *All Tomorrow's Parties.* New York: Ace Books.

Gilder, George. 1993. *Wealth and Poverty,* 2d ed. Oakland, Calif.: Institute for Contemporary Studies.

Girard, René. 1979. *Violence and the Sacred.* Trans. Patrick Gregory. Baltimore: Johns Hopkins University Press.

Gould, Stephen J. 1978. "Sociobiology: The Art of Storytelling." *New Scientist* 80: 530–33.

Guattari, Félix. 1989. *Cartographies Schizoanalytiques*. Paris: Galilée.

Haraway, Donna J. 1991. *Simians, Cyborgs, and Women: The Reinvention of Nature*. New York: Routledge.

Hardt, Michael, and Antonio Negri. 2001. *Empire*. Cambridge: Harvard University Press.

Harvey, David. 1990. *The Condition of Postmodernity*. Malden, Mass.: Blackwell.

Hayes, Charles, ed. 2000. *Tripping: An Anthology of True-Life Psychedelic Adventures*. New York: Penguin Compass.

Hayles, N. Katherine. 1999. *How We Became Posthuman: Virtual Bodies in Cybernetics, Literature, and Informatics*. Chicago: University of Chicago Press.

Heidegger, Martin. 1996. *Being and Time*. Trans. Joan Stambaugh. Albany: State University of New York Press.

Heim, Kristi. 2000. "Microsoft Showcases Home Wired to Internet." *Silicon Valley News*, 25 July. www0.mercurycenter.com/svtech/ news/indepth/docs/mshome072600.htm. Accessed April 2001.

Heim, Michael. 1993. *The Metaphysics of Virtual Reality*. New York: Oxford University Press.

Hérail, Rene James, and Edwin A. Lovatt. 1984. *Dictionary of Modern Colloquial French*. New York: Routledge.

Higgins, David. 2001. "Smile, Burglars, You're Starring on a Website." *The Sydney Morning Herald*, 9 February. www.smh.com.au/ news/0102/09/pageone/pageone1.html. Accessed April 2001.

Humphrey, Nicholas. 1999. *A History of the Mind*. New York: Copernicus Books.

Illegal Art. 1998. *Deconstructing Beck*.

Infesticons. 2000. *Gun Hill Road*. Big Dada Recordings.

Ivanov, Vladimir N. 1997. "Microbial Neural Network: Artificial Intelligence from Fungi." October. www.montegen.com/html/ body_a_multifunctional_biochip.htm. Accessed April 2001.

Jackson, Pamela. 1999. *The World Philip K. Dick Made*. Ph.D. diss., University of California, Berkeley.

James, Darius. 1995. *That's Blaxploitation!* New York: St. Martin's Griffin.

Jameson, Fredric. 1991. *Postmodernism; or, The Cultural Logic of Late Capitalism*. Durham, N.C.: Duke University Press.

Jeter, K. W. 1998. *Noir*. New York: Bantam.

Jurvetson, Steven T., and Tim Draper. 1999. "Viral Marketing." www.dfj.com/viralmarketing.html. Accessed April 2001.

Kant, Immanuel. 1987. *The Critique of Judgment*. Trans. Werner S. Pluhar. Indianapolis: Hackett.

———. 1996. *The Critique of Pure Reason*. Trans. Werner S. Pluhar. Indianapolis: Hackett.

Kauffman, Sturat. 1995. *At Home in the Universe: The Search for the Laws of Self-Organization and Complexity*. New York: Oxford University Press.

Kelly, Kevin. 1994. *Out of Control*. New York: Addison-Wesley.

Kharif, Olga. 2001. "Space Veggies, Phone Home." *Business Week Online*, 7 February. www.businessweek.com/bwdaily/dnflash/feb2001/nf2001027_666.htm. Accessed April 2001.

Kittler, Friedrich A. 1997. *Literature, Media, Information Systems*. Ed. John Johnston. Amsterdam: G + B Arts International.

Kuhn, Cynthia, Scott Swartzwelder, and Wilkie Wilson. 1998. *Buzzed: The Straight Facts about the Most Used and Abused Drugs from Alcohol to Ecstasy*. New York: W. W. Norton and Company.

Kumar, Aparna. 2001. "Is the FCC Quelling Free Speech?" *Wired News*, 18 April. www.wired.com/news/politics/0,1283,43078,00.html. Accessed April 2001.

Kurzweil, Ray. 2000. *The Age of Spiritual Machines*. New York: Penguin.

———. 2001. *The Law of Accelerating Returns*. www.kurzweilai.net/articles/art0134.html?printable=1. Accessed April 2001.

Lacan, Jacques. 1998. *The Four Fundamental Concepts of Psycho-Analysis*. Trans. Alan Sheridan. New York: W. W. Norton and Company.

Landsburg, Steven E. 1993. *The Armchair Economist*. New York: The Free Press.

Lanier, Jaron. 2001. "A Love Song for Napster." *Discover* 22: 2 (February). www.discover.com/feb_01/featnapster.html. Accessed April 2001.

Leibniz, G. W. 1973. *Monadology*. In *Leibniz, Philosophical Writings*, ed. G. H. R. Parkinson, trans. Mary Morris and G. H. R. Parkinson. Rutland, Vt.: Everyman's Library.

Lessig, Lawrence. 1999. *Code and Other Laws of Cyberspace*. New York: Basic Books.

Lettice, John. 2001. "Welcome to .Net: How MS Plans to Dominate Digital Music Sales." *The Register*, 15 February. www.theregister.co.uk/content/4/16959.html. Accessed April 2001.

Lévi-Strauss, Claude. 1963. *Structural Anthropology.* New York: Basic Books.

Levy, Stephen. 2001. *Crypto: How the Code Rebels Beat the Government—Saving Privacy in the Digital Age.* New York: Viking.

Lewontin, Richard C., Steven Rose, and Leon J. Kamin. 1984. *Not in Our Genes.* New York: Pantheon.

Litman, Jessica. 2001. *Digital Copyright.* New York: Prometheus Books.

Lütticken, Sven. 2002. "The Art of Theft." *New Left Review* 13 (January/February): 89–104.

Lycaeum. 2001. "Memetics Publications on the Web." users.lycaeum.org/~sputnik/Memetics/. Accessed April 2001.

Lyotard, Jean-François. 1985. *The Postmodern Condition: A Report on Knowledge.* Trans. Brian Massumi. Minneapolis: University of Minnesota Press.

MacLeod, Ken. 1999. *The Cassini Division.* New York: Tor Books.

———. 2000a. *Cosmonaut Keep.* London: Orbit.

———. 2000b. *The Sky Road.* New York: Tor Books.

———. 2000c. *The Stone Canal.* New York: Tor Books.

———. 2001. *The Star Fraction.* New York: Tor Books.

Mallarmé, Stephane. 1945. *Oeuvres Complètes.* Ed. Henri Mondor and G. Jean-Aubri. Paris: Pléiade.

Manovich, Lev. 2001. *The Language of New Media.* Cambridge: MIT Press.

Margulis, Lynn, and Dorion Sagan. 1991. *Microcosmos: Four Billion Years of Evolution from Our Microbial Ancestors.* New York: Touchstone.

———. 2002. *Acquiring Genomes: A Theory of the Origins of Species.* New York: Basic Books.

Margulis, Lynn, and Karlene V. Schwartz. 1988. *Five Kingdoms: An Illustrated Guide to the Phyla of Life on Earth.* 2d ed. New York: W. H. Freeman.

Marx, Karl. 1992. *Capital.* Vol. 1. Trans. Ben Fowkes. New York: Penguin.

———. 1993a. *Capital.* Vol. 3. Trans. David Fernbach. New York: Penguin.

———. 1993b. *Capital.* Vol. 2. Trans. David Fernbach. New York: Penguin.

Marx, Karl, and Friedrich Engels. 1848. *The Manifesto of the Communist Party.* www.anu.edu.au/polsci/marx/classics/manifesto.html. Accessed April 2001.

Massumi, Brian. 2002. *Parables for the Virtual: Movement, Affect, Sensation.* Durham, N.C.: Duke University Press.

Mauss, Marcel. 1967. *The Gift.* Trans. Ian Cunnison. New York: W. W. Norton and Company.

May, Tim C. 2001. "True Nyms and Crypto Anarchy." In Vernor Vinge, *True Names and the Opening of the Cyberspace Frontier*, ed. James Frenkel. New York: Tor Books.

McKenna, Terence. 1983. "Tryptamine Hallucinogens and Consciousness." www.deoxy.org/t_thc.htm. Accessed April 2001.

———. 1984. "New Maps of Hyperspace." www.deoxy.org/t_newmap.htm. Accessed April 2001.

———. 1991. "Plan Plant Planet." www.deoxy.org/t_ppp.htm. Accessed April 2001.

———. 1992. *The Archaic Revival.* New York: Harper.

———. 1994. "Timewave Zero." www.spiritweb.org/Spirit/timewave-zero.html. Accessed April 2001.

McLeod, Kembrew. 2001. *Owning Culture: Authorship, Ownership, and Intellectual Property Law.* New York: Peter Lang.

McLuhan, Marshall. 1994. *Understanding Media: The Extensions of Man.* Cambridge: MIT Press.

McLuhan, Marshall, and Quentin Fiore. 1961. *War and Peace in the Global Village.* New York: Bantam.

———. 1967. *The Medium Is the Massage.* New York: Bantam.

McMenamin, Mark A. S. 1998. *The Garden of Ediacara.* New York: Columbia University Press.

Miéville, China. 2000. *Perdido Street Station.* New York: Del Ray/Ballantine Books.

Miller, George A., et al. 2001. *WordNet: A Lexical Database for English.* www.cogsci.princeton.edu/~wn/. Accessed April 2001.

Misha. 1999. *Red Spider, White Web.* La Grande: Wordcraft of Oregon.

Moravec, Hans. 1990. *Mind Children.* Cambridge: Harvard University Press.

———. 2000. *Robot: Mere Machine to Transcendent Mind.* New York: Oxford University Press.

Munroe, Jim. 2002. *Everyone in Silico.* Toronto: No Media Kings.

Narby, Jeremy. 1999. *Cosmic Serpent: DNA and the Origins of Knowledge.* New York: J. P. Tarcher.

Nietzsche, Friedrich. 1968a. *Twilight of the Idols/The Antichrist.* Trans. R. J. Hollingdale. New York: Penguin.

————. 1968b. *The Will to Power.* Trans. Walter Kaufmann and R. J. Hollingdale. New York: Vintage.

————. 1969. *On the Genealogy of Morals/Ecce Homo.* Trans. Walter Kaufman. New York: Vintage.

————. 1974. *The Gay Science.* Trans. Walter Kaufmann. New York: Vintage.

Noon, Jeff. 1997. *Nymphomation.* London: Corgi Books.

————. 1999. *Needle in the Groove.* London: Transworld Publishers.

Ortony, Anthony. 2000. *Cognitive Science: Foundations of the Learning Sciences.* Syllabus for Learning Sciences 403, Northwestern University, Autumn 2000. www.ls.sesp.nwu.edu/courses/403/. Accessed April 2001.

Oss, O. T., and O. N. Oeric. 1992. *Psilocybin: Magic Mushroom Grower's Guide.* Quick American Archives. www.deoxy.org/mushword.htm. Accessed April 2001.

Peckham, Morse. 1988. *Explanation and Power: The Control of Human Behavior.* Minneapolis: University of Minnesota Press.

Pellerin, Cheryl. 1998. *Trips: How Hallucinogens Work in Your Brain.* New York: Seven Stories.

Penrose, Roger. 1991. *The Emperor's New Mind.* New York: Penguin.

Peters, Tom. 1994. *The Tom Peters Seminar: Crazy Times Call for Crazy Organizations.* New York: Vintage.

Pinker, Steven. 1997. *How the Mind Works.* New York: W. W. Norton and Company.

Pinker, Steven, and Robert Wright. 2000. "An Email Conversation about Robert Wright's Nonzero." *Slate Magazine,* February. slate.msn.com/default.aspx?id=2000143. Accessed April 2001.

Platt, Charles. 2000. "Streaming Video." *Wired* 8: 11 (November). www.wired.com/wired/archive/8.11/luvvy_pr.html. Accessed April 2001.

Prigogine, Ilya, and Isabelle Stengers. 1984. *Order out of Chaos: Man's New Dialogue with Nature.* New York: Bantam.

Putnam, Robert D. 2001. *Bowling Alone: The Collapse and Revival of American Community.* New York: Touchstone Books.

Pynchon, Thomas. 1973. *Gravity's Rainbow.* New York: Viking.

————. 1991. *Vineland.* New York: Penguin.

Regan, Ciaran. 2001. *Intoxicating Minds: How Drugs Work.* New York: Columbia University Press.

Rorty, Richard. 1979. *Philosophy and the Mirror of Nature.* Princeton: Princeton University Press.

Rose, Tricia. 1994. *Black Noise: Rap Music and Black Culture in Contemporary America.* Hanover: Wesleyan University Press.

Ross, Andrew. 2000. *The Celebration Chronicles: Life, Liberty, and the Pursuit of Property Value in Disney's New Town.* New York: Ballantine Books.

Safire, William. 1997. "The Misrule of Law: Singapore's Legal Racket." *New York Times,* 1 June. www.gn.apc.org/sfd/Link% 20Pages/Link%20Folders/The%20Law/misrulelaw.html. Accessed April 2001.

Sanders, Edmund. 2001. "Corporate Free Speech Battle Is Escalating." *Los Angeles Times,* 27 May. www.latimes.com/business/ 20010527/t000044302.html. Accessed May 2001.

Sansoni, Silvia. 1999. "Word-of-Modem." *Forbes Magazine,* 5 July. www.forbes.com/forbes/1999/0705/6401118a.html. Accessed April 2001.

Sardar, Ziauddin. 2001. "Trapped in the Human Zoo." *New Statesman,* 19 March. www.consider.net/forum_new.php3?new Template=OpenObject&newTop=20010390016&new DisplayURN=200103190016. Accessed April 2001.

Saroyan, Aram. 1979. *Genesis Angels: The Saga of Lew Welch and the Beat Generation.* New York: William Morrow.

Scarbrough, William. 1999. *Prosthetic.* CD-ROM.

Schumpeter, Joseph. 1984. *Capitalism, Socialism, and Democracy.* New York: Harper.

Searle, John. 1997. *The Mystery of Consciousness.* New York: New York Review of Books Press.

Shaviro, Steven. 1997. *Doom Patrols: A Theoretical Fiction about Postmodernism.* New York: Serpent's Tail.

———. 1998. "Deconstructing Beck." *Artbyte* 1, no. 2 (July/August): 16–17.

———. 1999a. "Through a Monitor, Darkly: William Scarbrough's *Prosthetic.*" *Artbyte* 2, no. 3 (September/October): 26–27.

———. 1999b. "Virtual Pleasure, Virtual Pain: Diane Gromala's *Dancing with the Virtual Dervish.*" *Artbyte* 2, no. 1 (April/May): 20–21.

———. 2001. "Regimes of Vision: Kathryn Bigelow's *Strange Days.*" *Polygraph* 13: 59–68.

———. 2002a. *Stranded in the Jungle.* http://www.shaviro.com/ Stranded/. Accessed April 2002.

———. 2002b. "The Erotic Life of Machines," in *parallax* 25 (October–December 2002), 21–31.

Siegfried, Tom. 2000. *The Bit and the Pendulum*. Hoboken: John Wiley and Sons.

Smith, Patrick S. 1986. *Andy Warhol's Art and Films*. Ann Arbor: UMI.

Spicer, Jack. 1975. *The Collected Books of Jack Spicer*. Edited and with a commentary by Robin Blaser. Los Angeles: Black Sparrow Press.

———. 1998. *The House That Jack Built: The Collected Lectures of Jack Spicer*. Ed. Peter Gizzi. Hanover: Wesleyan University Press.

Stafford-Fraser, Quentin. 1995. *The Trojan Room Coffee Pot: A (Non-Technical) Biography*. www.cl.cam.ac.uk/coffee/qsf/coffee.html. Accessed April 2001.

Stamler, Bernard. 2001. "If You Hate Web Ads, You Can 'Just Say No.'" *New York Times*, 13 June. www.nytimes.com/2001/06/13/technology/13STAM.html. Accessed June 2001.

Standage, Tom. 1999. *The Victorian Internet*. New York: Berkley Books.

Stephenson, Neal. 1992. *Snow Crash*. New York: Bantam Books.

Sterling, Bruce. 1999. *Distraction*. New York: Bantam Books.

Strassman, Rick. 2001. *DMT: The Spirit Molecule*. Vermont: Park Street Press.

Sullivan, Andrew. 2000. "The He Hormone." *New York Times Magazine*, 2 April, 46ff.

Surveillance Camera Players. 1995. "Guerilla Programming of Video Surveillance Equipment." www.notbored.org/gpvse.html. Accessed April 2001.

———. 1996. "Programming Note for First Season, November 1996." www.notbored.org/scp.html. Accessed April 2001.

———. 1999. "Time in the Shadows of Anonymity." www.notbored.org/transparent.html. Accessed April 2001.

Taussig, Michael. 1992. *Mimesis and Alterity: A Particular History of the Senses*. New York: Routledge.

Teilhard de Chardin, Pierre. 1975. *The Phenomenon of Man*. Trans. Bernard Wall. New York: Harper and Row.

Thornhill, Randy, and Craig T. Palmer. 2000. *A Natural History of Rape: Biological Bases of Sexual Coercion*. Cambridge: MIT Press.

Vinge, Vernor. 1993. *The Coming Technological Singularity: How to Survive in the Post-Human Era*. www-rohan.sdsu.edu/faculty/vinge/misc/singularity.html. Accessed April 2001.

Waggoner, Ben, and Brian Speer. 1998. *Introduction to the Fungi*.

University of California, Berkeley. www.ucmp.berkeley.edu/ fungi/fungilh.html. Accessed April 2001.

Warhol, Andy. 1975. *The Philosophy of Andy Warhol.* New York: Harvest/HBJ.

———. 1989. *The Andy Warhol Diaries.* Ed. Pat Hackett. New York: Warner Books.

Weber, Max. 1958. *The Protestant Ethic and the Spirit of Capitalism.* Trans. Talcott Parsons. New York: Scribner.

Whatis.com. 2001. "Viral Marketing." searchcrm.techtarget.com/ sDefinition/0,,sid11_gci213514,00.html. Accessed April 2001.

Whyte, William H. 1956. *The Organization Man.* New York: Simon and Schuster.

Willmott, Don. 1999. "The Joys of Viral Marketing." *PC Magazine,* 28 July. www.zdnet.com/pcmag/stories/opinions/0,7802,2298284, 00.html. Accessed April 2001.

Wilson, Edward O. 1998. *Consilience: The Unity of Knowledge.* New York: Alfred A. Knopf.

Wittgenstein, Ludwig. 1999. *Philosophical Investigations.* 3d ed. Trans. G. E. M. Anscombe. Upper Saddle River, N.J.: Prentice-Hall.

Wright, Robert. 2001. *Nonzero: The Logic of Human Destiny.* New York: Vintage.

Zahavi, Amotz, and Avishag Zahavi. 1997. *The Handicap Principle.* New York: Oxford University Press.

Zimmer, Carl. 2000. *Parasite Rex.* New York: The Free Press.

Žižek, Slavoj. 1999. "*The Matrix;* or, The Two Sides of Perversion." on1.zkm.de/netCondition/matrix/zizek.html. Accessed April 2001.

———. 2001. *On Belief.* New York: Routledge.

Filmography

Cronenberg, David. 1983. *Videodrome*.

Cunningham, Chris, and Björk. 1999. *All Is Full of Love*. Elektra DVD.

Cunningham, Chris, and R. D. James. 1998. "Come to Daddy," in *Come to Viddy*. WEA/London/Sire. VHS tape.

Ferrara, Abel. 1998. *New Rose Hotel*.

Godard, Jean-Luc. 1965. *Alphaville*.

Ra, Sun. 1972. *Space Is the Place*.

Roeg, Nicholas. 1976. *The Man Who Fell to Earth*.

Romero, George. 1978. *Dawn of the Dead*.

Scott, Ridley. 1982. *Blade Runner*.

Wachowski Brothers. 1999. *The Matrix*.

Warhol, Andy. 1964. *Empire*.

Index

Steven Shaviro teaches cinema studies and English at the University of Washington. He is the author of *The Cinematic Body* (Minnesota, 1993), *Doom Patrols: A Theoretical Fiction about Postmodernism,* and *Passion and Excess: Blanchot, Bataille, and Literary Theory,* as well as numerous articles about film and contemporary culture.